WEATHER INFERENCE
FOR BEGINNERS

STORM CLOUD DEVELOPING OVER LONDON: 10.30 G.M.T. ON 5.9.36

(see page 34)

WEATHER INFERENCE
FOR BEGINNERS

MADE CLEAR IN A SERIES OF
ACTUAL EXAMPLES

BY

D. J. HOLLAND, M.A.

CAMBRIDGE
AT THE UNIVERSITY PRESS
1953

CAMBRIDGE UNIVERSITY PRESS
Cambridge, New York, Melbourne, Madrid, Cape Town,
Singapore, São Paulo, Delhi, Mexico City

Cambridge University Press
The Edinburgh Building, Cambridge CB2 8RU, UK

Published in the United States of America by Cambridge University Press, New York

www.cambridge.org
Information on this title: www.cambridge.org/9781107619494

First published 1953
First paperback edition 2013

A catalogue record for this publication is available from the British Library

ISBN 978-1-107-61949-4 Paperback

To
MY MOTHER

CONTENTS

CONTENTS

ILLUSTRATIONS

ILLUSTRATIONS

PREFACE

This book is for all who are interested in their own daily weather.

It is a professional meteorologist's running commentary on the weather which he himself had already observed and recorded ten years earlier as a schoolboy when he had no more facilities than have since become available to everybody. Having thus encouraged the reader, its purpose is to illustrate the analysis and forecasting of weather with local observations and charts from the Meteorological Office.

For readers unfamiliar with weather theory the first chapter just outlines its ideas. For those who are also unfamiliar with Meteorological Office procedure the second chapter explains how (for brevity) all the reports are coded. The coded weather reports are then listed with running commentaries explaining the weather analysis. Even with only one season (one autumn) the main ideas of the subject are covered. Difficult ones for beginners are purposely explained with the least possible technical or mathematical language. The air's elementary thermodynamics, tephigrams and radiation, for instance, are summed up in an unusual way, after which the account of its hydrodynamics or wind theory also includes an original working design, while ideas of simplified charts and classification of weather types are put forward. Another original feature is that the illustrations are not specially selected, but are taken day by day as they come.

After the year 1948 the standard international codes were altered. Readers using Meteorological Office *Daily Weather Reports* (abbreviated to *D.W.R.*) will now find the new codes in other Air Ministry Meteorological Office publications sold by H.M. Stationery Office. The weather reports in this book have accordingly all been recoded. But as the writer, like many readers, originally had no better guide than *D.W.R. Introductions*, which were not really a guide but just a short dictionary to the codes, he unwittingly departed slightly from standard practice. As the old codes could not be exactly translated into the new ones either, he cannot attempt in this book to stick to the new codes exactly, but explains the codes he prefers to use to make the weather reports as clear to the reader as possible.

Although not reproduced throughout the book, many charts copied from Air Ministry *D.W.R.* are included to illustrate weather-map types.

PREFACE

To quote the book's own conclusion, not only may the general reader learn something of what lies behind the Meteorological Office forecasts and inferences and how far he may make his own from *Daily Weather Reports*, but other beginners in the Meteorological Office itself who already know how to make weather charts will have a better idea how to use them.

Acknowledgements are due to the staff of the Cambridge University Press for valuable advice and assistance in the preparation of the book, to the Director, Meteorological Office, Air Ministry, London, and to the Controller of H.M. Stationery Office for permission to include extracts from official publications.

D. J. H.

I

FIRST PRINCIPLES

Every day you either make or receive an *inference* of your own past, present or future weather. 'An anticyclone', you read in the paper, 'is centred over this country.' 'Turned out nice again,' you say when you see the sun.

Those are inferences. In this country you cannot do without them. Upon them may hang all your plans for the day. On their accuracy rests success or disaster, whether to health or to clothes, to picnics or harvests, work or play. You need to know both how to interpret the inferences you receive, and how to make your own more easily.

The idea of this book is to help you. It is a single observer's record of his own weather with a running commentary to explain all its changes. Whilst requiring no more elaborate instrument than a thermometer, it is quite good enough to illustrate how meteorologists in this country think of the weather, how they talk of it, how they have it reported, charted, analysed and finally forecast.

What we must learn is their language. They think of the weather in rather the same way as doctors think of your health. They *diagnose* and *prognose* it. First of all they have it reported. The report is not the whole story. It just says what happens on the surface. For the doctor it may be a rash on the surface of your body. For the meteorologist it may be rain falling on the surface of the earth. It is just the chief *symptom*. Your doctor also notes exactly your temperature and perhaps blood pressure. So do we note the temperature and barometric pressure of the air. Just as an abnormal body temperature means something wrong or abnormal inside, so an abnormal barometric pressure means something abnormal aloft. You can no more easily go up to see for yourself than the doctor can see for himself inside you. Without an operation you must learn to judge from outside. That is diagnosis. The simplest observations, in fact, are not actual measurements but mere classifications of how things look, or how you feel. The expert completes the picture, and uses the picture to size things up. His diagnosis sometimes is only of the immediate cause of the symptoms. At other times it goes deeper. Diagnosis, of course, is half-way to prognosis or forecast. Only meteorologists cannot cure! This book is of observations with diagnoses, showing how they were made for one place for one season.

For diagnosis the doctor must know how the body is built. So must the meteorologist know the atmosphere. How is the atmosphere built? It is really an ocean. It contains water, but mostly as vapour. Rain, snow, clouds, fogs—all these kinds of bad weather consist of water in liquid or solid form. As long as the water is only vapour the weather is fine. So the weather is simply a matter of water changing its state. But that means gain or loss of latent heat, for which Nature has a rationing system. The pressure and temperature determine a limit to the amount of water vapour the air can hold. Suppose they are lowered too far for the air any longer to hold all its water as vapour. The surplus water must be condensed. Into what? Into fog or clouds or even into PRECIPITATION as rain or snow. That is bad weather. The weather, you see, consisting of

$$\left.\begin{array}{l} \textit{precipitation} \\ \textit{clouds} \\ \textit{visibility} \\ \textit{wind} \end{array}\right\} \text{familiar, non-technical elements}$$

is determined by laws of Nature with

$$\left.\begin{array}{l} \textit{temperature} \\ \textit{pressure} \\ \textit{moisture} \end{array}\right\} \text{technical elements.}$$

Having reported them, what do we do with them next? The best way here to use what is known, in order to estimate what has hitherto been unknown, is to plot them on charts and then draw the charts up. In experienced hands this is a remarkably rapid and accurate method, combining science with art. In effect the weather consists of certain *properties* of the air. Air is a fluid. Fluid motion has its mechanics as well as its beauty, its science as well as its art. This particular science is hydrodynamics, and has its own technical terms. Any change in any property of a particular MOVING PARTICLE or ELEMENT of a fluid is called a TOTAL change. Any change in the property, on the other hand, at a particular FIXED PLACE in the fluid is called a LOCAL change. Air which rises above hot ground, for instance, cools as it rises: that is a 'total' change of temperature in the rising air. But at the ground, a *fixed place* in the rising air, the temperature may stay the same all the time. The 'local' change is zero. A change of weather, then, at a fixed place, is simply a local change in certain properties of moving air. Obviously it can be analysed into two parts:

I. 'Total' change;

II. 'Advection', or BODILY TRANSPORT of properties by the flow.

To the weather, therefore,

total change contributes NEW properties of OLD air, while
advection contributes OLD properties of NEW air.

For total change the appropriate chart is ideally a map of the weather at all levels, one time and one place. That is an aerological or upper-air chart. For advection, on the other hand, the appropriate chart is ideally a map of the weather at all places, one time and one level. That is a synoptic chart.

SYNOPTIC weather analysis, in short, tells you *where* and *what* the air is, while AEROLOGICAL analysis tells you *how* or in what state it is. The atmosphere consists of great masses of air, in each of which many properties are horizontally almost uniform, so that a new type of weather occurs with a new air mass. The line or zone of transition is called a front. So the weather is commonly analysed in terms of WEATHER TYPES, AIR MASSES and FRONTS.

Your doctor must also know how your body *works*, and so must we know the atmosphere. How does it work? Where is the air hottest? In the tropics. So there it must rise, accordingly flowing in over the ground and out up aloft. Beneath it the world is turning round. But motion is purely relative. We think of the ground as being at rest while the atmosphere does the turning. In northern latitudes as the earth turns anti-clockwise relative to the heavens, so must the air or heavens turn clockwise or always *to the right* relatively to the earth. The air, in short, when moving freely over the earth, always starts to turn to the right as if pushed by a force. Otherwise we say it is not moving freely. At any rate we always assume this force. From high to low pressure, of course, is also a force. To balance it, therefore, at any one level, the push to the right must act from low to high pressure. So keeping the high pressure always upon the right-hand side of its path the wind must blow *round* the high or low pressure according to the famous law of Buys Ballot: 'If (in northern latitudes) you stand with your back to the wind, then the pressure is lower upon your left hand than upon your right.'

Winds thus blow into the tropics not from north, but from a more easterly quarter. In southern latitudes, where the winds turn to the left instead, they enter the tropics not from south but again from an easterly quarter. Emerging therefore aloft from a westerly quarter in both hemispheres they presently come down to earth as ex-tropical WESTERLIES, such as the old mariners' Roaring Forties, Furious Fifties and Shrieking Sixties. The high pressure, which is then on the right-hand side in northern latitudes or the left-hand in southern latitudes, must in either case be upon the equator side. As it must be *between* the Forties and the tropical easterlies, it must be at the *edge* of the tropics. So there must be a SUBTROPICAL HIGH-PRESSURE belt. On the other side of the westerlies must be low pressure.

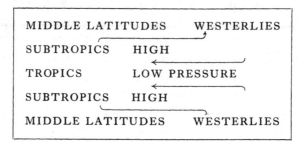

Fig. 1. General circulation.

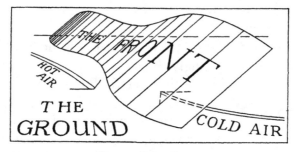

Fig. 2. Polar front.

Should the pressure be higher again beyond it towards the poles, then there must be polar easterly winds. Between these and the warmer westerlies is a kind of war. The battle front is the so-called POLAR FRONT.

What is it like? The force of gravity, as you know, makes hot air tend to lie horizontally over cold. The extra force that pushes the air to the right or left makes it lie at a slant instead. The mechanics of it are not very simple, but that is the rough idea. The 'front' is therefore a sloping surface between different air streams which you may regard as two kinds of fluid with different flow. Now when wind blows over water we also have two kinds of fluid which differ in flow: and what does the surface between them do then? As long as the two streams differ enough there are WAVES. Either

4

short ripples or long waves, once started, they grow by themselves, and as they travel they may do damage. So does the polar front. So, in fact, does the front or boundary surface between *any* two sufficiently different air masses. We are most concerned with its big long waves that develop DEPRESSIONS, extensive bad weather and immense power.

Under water we can make the pressure uneven at any one level by having some hot and some cold, as we do in the bath. In the same sort of way we have average pressure low over land in the summer and high in the winter. But in the bath to get really low pressure you pull out the plug. Quiet at first, the flow is nevertheless unstable, as it may change to a whirl. Quiet frontal air-flow may likewise rapidly change to a CYCLONE or circulation round a low-pressure centre, tending to move with the warmer air until it has so distorted the original wave as to have destroyed it. Others follow,

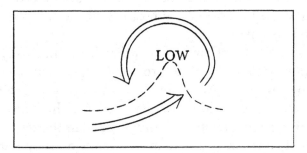

Fig. 3. Air circulation round a wave in a front.

often successively farther away from the poles until the cold air breaks right through the region to reign in peace with high pressure. Victorious in its final battle it sweeps all the old air away to the tropics and settles itself in sole occupation. Surging into the tropics, in fact, it feeds the sub-tropical high-pressure zone with its fresh high pressure, so that it may make its influence felt all the way to the equator. Meanwhile, of course, being strongly heated over the tropics it becomes tropical air and rises well laden with moisture over tropical seas to return to a new polar front. So the rest of the world gets its rain.

That is the shortest possible outline of HOW THE AIR-FLOW ALTERS THE WEATHER BY BRINGING OLD PROPERTIES OF NEW AIR. Now what, on the other hand, brings *new* properties into *old* air?

Air is HEATED by (1) descent and compression,
          (2) warmer land or sea,
          (3) condensation of water.

5

It is COOLED by (1) ascent and expansion,
                   (2) colder land or sea,
                   (3) evaporation of water.

Sufficient ascent of air over fronts or mountains, for instance, makes water vapour condense into clouds or even rain. We also say that hot air rises. If it is hotter than the air *beside* it, that is true enough. But if merely hotter than the air *above*, then it does not necessarily rise. For in rising it cools itself at a standard rate, for which we have to allow. It is only *potentially* hot air that freely rises. This free rising is CONVECTION, which always goes on in the air over warmer land or sea. Any clouds formed by it tend to be of the puffy CUMULUS type. Any rain or snow from them tends to be of the SHOWERY type. Inland they normally only develop by day when the earth is warmer than all the air, but at sea (and therefore also on windward coasts) they may go on almost equally well day and night until a warmer air mass arrives.

Descent or SUBSIDENCE of air, on the other hand, normally has to make up for horizontal outflow over the ground or sea, such as in a region of high or of rising pressure. At sea-level itself, of course, the descent must stop, so the air aloft is warmed more than the air at the bottom. This tends to raise the upper-air temperature even higher than the temperature of the air underneath. The result is called an INVERSION or temperature. 'Inversion', in fact, means not merely this abnormal state of affairs but the actual layer of air that is warmer above than below. It is important because, like a kind of ceiling, it stops air rising. In clear air it cannot be seen, but in hazy air it appears as a haze top, while in damp air it is the top of a layer o cloud. Similar in effect is *any* layer of air *potentially* warmer above than below. If too low for clouds to be formed below it at all, it must tend to make the weather FINE (unless there are clouds higher up). If the air is damp enough to make clouds below it, the weather will be of the CLOUDY or overcast type. If the layer is higher, however, so that the clouds have room to begin to dry out, the weather may be of the FAIR type. There you see, are three main weather types already partly explained.

Having outlined it, we must consider what weather analysis really is. To-morrow's weather, for instance, is forecast from to-day's. How? Ideally by applying all the laws of Nature to all to-day's weather conditions. In practice we may know more of the laws than of the conditions, or *vice versa*, so that we just have to do our best. THEORY tells us what laws to apply, while OBSERVATIONS tell us to-day's conditions. The idea of weather analysis is to express them in the same language; for only then can the laws be applied. What is the language to be?

Our analysis of total change, namely of the way in which properties change as the air goes along, is largely a matter of thermodynamics, the theory of heat flow. That is part of physics. It speaks of temperature, moisture, pressure—physical terms. Analysis, on the other hand, of advection, namely of the way in which properties are bodily carried along by the air-flow, is largely a matter of aerodynamics which is a branch of hydrodynamics, the theory of fluid flow. That is part of mechanics. It speaks of density, pressure and circulation or flow—rather more mathematical terms. Air density, however, can be expressed in terms of pressure and moisture and temperature, circulation or flow being a matter of wind. *Both* parts of weather analysis are thus boiled down to the same elements, MOISTURE, PRESSURE, TEMPERATURE and WIND—the ABCD of the language. If theory and practice are to talk the same language, then we must express observations in the same terms as these. Wind is already one of the four weather elements we have observed. Moisture, pressure and temperature can equally well be measured. All that remain are PRECIPITATION, CLOUDS and VISIBILITY to be described in figures too.

II

THE CODES

Synoptic weather reports describe

    (i) *precipitation or general weather*
    (ii) *clouds* — capable of adequate description
    (iii) *visibility* — in figures by mere classification,
    (iv) *wind* — requiring no instruments,

    (v) *temperature*
    (vi) *pressure* — requiring instruments.
    (vii) *moisture*

Our reports (see e.g. pp. 22, 23) use (i)–(v), tabulated in thirteen columns. Instruments are often required elsewhere for some of the first four elements, particularly for wind, but have not been used for reports in this book.

Column 1 gives the Observation Station Number. The weather partly depends on PLACE, which must therefore always be stated, preferably by a simple number. The following table explains the place numbers used in this book.

No.	Name	Latitude N	Longitude W	Height (ft.)
01	West Norwood, London	51 26	00 06	270
02	Streatham Common	51 25	00 07	250
03	Merstham, Surrey	51 17	00 09	400
04	Norbury	51 24	00 07	100
05	Wandsworth	51 27	00 12	100
06	Mortlake	51 28	00 16	20
07	Kew	51 29	00 17	20
08	Upton, Hants.	51 17	01 29	400
09	Andover	51 12	01 29	250
10	Brookwood	51 18	00 38	150
11	Tooting	51 26	00 10	50
12	Wimbledon	51 25	00 12	100
13	Malden	51 24	00 15	50
14	Kingston	51 24	00 17	40
15	Hampton Court	51 24	00 20	30
16	Mitcham	51 24	00 10	100
17	Addiscombe	51 23	00 04	150
18	West Dulwich	51 26	00 05	125
19	Upper Norwood	51 25	00 05	350
20	London (the City)	51 30	00 05	20

They were assigned by the writer himself, as you yourself might, starting at home with No. 1 and then adding one to the list every time a change of weather was noted from some place not already numbered. All but a few observations, however, were made near London, so that the changes of place do not make much difference. The *idea* of numbering observation stations, however, is universal.

Column 2 gives the DATE.

Column 3 gives the TIME, G.M.T. This book's reports, unlike standard Meteorological Office ones, were made not at fixed hours but just whenever changes were noted. That seemed the best way, though restricted by other duties as well as by every night's sleep. If your reports are made for a central office with hundreds of others to add to the picture, then naturally you report at regular times; but to complete the picture yourself you note the weather just whenever you can. That is what the writer has done.

Column 4 describes the WEATHER by standard abbreviations, thus:

b=blue sky: fine weather
bc=blue sky with clouds: fair weather
c=cloudy
d=drizzle (moderate, or unspecified)
dd=drizzle (moderate), continuous
d_0=drizzle (slight)
d_0d_0=drizzle (slight), continuous
D=drizzle (heavy)
DD=drizzle (heavy), continuous
e=wet air, wet fog
f=fog
F=dense fog
g=gale
h=hail
i=intermittent...
j=...in sight, though not occurring at your place itself
k=storm of...(e.g. sand) arising with violent wind

l=lightning
m=mist
o=overcast
p=passing showers
q=squalls
r=rain
rr=continuous moderate rain, and so on as for drizzle
s=snow
ss=continuous moderate snow, and so on
t=thunder
u=ugly, threatening sky
v=unusual visibility
w=dew
x=frost
y=very dry air
z=haze

Capital letters thus mean a heavy or intense variety; suffix '$_0$' means 'slight'; repetition denotes continuity, while prefix 'i' stands for intermittence. '+' denotes increase, '−' means decrease, and (in this book only) '>' means 'becoming'. '/' (oblique stroke), followed by any weather abbreviation, means that that weather has arisen *since* the time of the last recorded observation although it may now have ceased. SHOWERS are distinguished from 'intermittent' or 'occasional' rain or snow, not because they are any less wet, but because they come from quite different upper-air types, with different causes, different accompaniment such as bright intervals or squalls, and different development, such as a tendency to die out inland at night.

Column 5 gives the total CLOUD amount in OKTAS or eighths of the sky (as seen from the ground).

Columns 6–8 describe the forms or types of clouds, low (L.), medium (M.), and high (H.). Notoriously hard for beginners to describe correctly, they are nevertheless our most valuable information. 'Show me your tongue' says the doctor; or 'Let me see your rash'. A glance at those outward signs is often enough for his diagnosis. Likewise a glance at the clouds is often enough for a diagnosis of how all the upper air is, if not indeed to tell what type it is and whence it has come. Just as the doctor looks only for certain signs as symptoms of certain disorders, so we classify clouds by certain features as symptoms of certain weather types. Readers are recommended to learn the standard international classification illustrated in a cloud atlas, of which a useful summary is the Air Ministry's Meteorological Office handbook of *Cloud Forms* (H.M.S.O., 1949 edition), from which some definitions will now be quoted as well as those used by the writer. The writer himself had originally no more guidance than the bare definitions published in the quarterly *Introduction* to Meteorological Office *Daily Weather Reports*, so that again he departed slightly from Meteorological Office standards, and these in turn have since been slightly altered, with the result that the cloud-type numbers in this book do not quite correspond to those you will nowadays find in *Daily Weather Reports*. As far as possible, however, they have been brought up to date. Here are the main types:

HIGH CLOUDS
(GENERALLY ABOVE 6 KM. OR 20,000 FT.)

1. CIRRUS (abbreviated to CI), the wispy type made of crystals of ice, with blue sky showing.

Officially defined as

DETACHED CLOUDS OF DELICATE AND FIBROUS APPEARANCE, WITHOUT SHADING, GENERALLY WHITE IN COLOUR, OFTEN OF A SILKY APPEARANCE.

Cirrus appears 'in the most varied forms such as isolated tufts, lines drawn across a blue sky, branching feather-like plumes, curved lines ending in tufts, etc.; they are often arranged in bands which cross the sky like meridian lines, and which, owing to the effect of perspective, converge to a point on the horizon, or to two opposite points'.

2. CIRROCUMULUS (Cc).

A CIRRIFORM LAYER OR PATCH COMPOSED OF SMALL WHITE FLAKES OR OF VERY SMALL GLOBULAR MASSES, WITHOUT SHADOWS, WHICH ARE ARRANGED IN GROUPS OR LINES, OR MORE OFTEN IN RIPPLES RESEMBLING THOSE OF THE SAND ON THE SEA SHORE.

Generally 'represents a degraded state of cirrus and cirrostratus both of which may change into it. In this case the changing patches often retain some fibrous structure in places. *Real cirrocumulus is uncommon.* It must not be confused with small altocumulus on the edges of altocumulus sheets.'

3. CIRROSTRATUS (Cs)

A THIN WHITISH VEIL, WHICH DOES NOT BLUR THE OUTLINES OF THE SUN OR MOON, BUT GIVES RISE TO HALOES.

Is 'sometimes quite diffuse and merely gives the sky a milky look; sometimes it more or less distinctly shows a fibrous structure with disordered filaments'.

MIDDLE OR MEDIUM CLOUDS
(GENERALLY AT 2–6 KM.)

4. ALTOCUMULUS (Ac), generally in layer form but broken, not quite continuous but made of drops of water instead of crystals of ice, and associated with the FAIR or SHOWERY or THUNDERY weather types, it is

A LAYER OR PATCHES), COMPOSED OF LAMINAE OR RATHER FLATTENED GLOBULAR MASSES, THE SMALLEST ELEMENTS OF THE REGULARLY ARRANGED LAYER BEING FAIRLY SMALL AND THIN, WITH OR WITHOUT SHADING.

'These elements are arranged in groups, in lines or waves, following one or two directions and are sometimes so close together that their edges join. The thin and translucent edges of the elements often show IRISATIONS which are rather characteristic of this class of cloud.'

5. ALTOSTRATUS (As), unbroken, continuous grey, watery-looking, sometimes so thick as to extend down below 2 Km., when it becomes NIMBOSTRATUS, this is a

STRIATED OR FIBROUS VEIL, MORE OR LESS GREY OR BLUISH IN COLOUR.

'This cloud is like thick cirrostratus but without halo phenomena; the sun or moon shows vaguely, with a faint gleam, as though through ground glass. Sometimes the sheet is thin with forms intermediate with cirro-stratus. Sometimes it is very thick and dark, sometimes even completely hiding the sun or moon. In this case differences of thickness may cause relatively light patches between very dark parts; but the surface never shows real relief, and the striated or fibrous structure is always seen in places in the body of the cloud.'

With this is associated the CONTINUOUS RAINY weather type, or the intermittent type of rain (or snow) which may not fall for as long as an hour at a time but always falls from continuous sheets of cloud, as opposed to SHOWERS which only come from isolated masses of cloud that do not cover the whole sky for long.

LOW CLOUDS
(GENERALLY BELOW 2 KM.)

6. STRATOCUMULUS (Sc), spreading or developing horizontally, when weather is of the CLOUDY type, is

A LAYER (OR PATCHES) COMPOSED OF GLOBULAR MASSES OR ROLLS; THE SMALLEST OF THE REGULARLY ARRANGED ELEMENTS ARE FAIRLY LARGE; THEY ARE SOFT AND GREY, WITH DARKER PARTS.

'These elements are arranged in groups, in lines, or in waves, aligned in one or two directions. Very often the rolls are so close that their edges join together; when they cover the whole sky they have a wavy appearance.'

7. STRATUS (St), also spreading, like fog above ground, when weather is of the OVERCAST type, is thus defined as

A UNIFORM LAYER OF CLOUD, RESEMBLING FOG, BUT NOT RESTING ON THE GROUND.

'When this very low layer is broken up into irregular shreds it is desig-nated FRACTOSTRATUS (FS).'

8. NIMBOSTRATUS (Ns) with weather of RAINY (or snow) type, is

A LOW, AMORPHOUS AND RAINY LAYER, OF A DARK GREY COLOUR AND NEARLY UNIFORM.

'It appears as though feebly illuminated seemingly from inside. When it gives precipitation this is in the form of continuous rain or snow. But precipitation alone is not sufficient criterion to distinguish the cloud which should be called nimbostratus even when no rain or snow falls from it.

There is often precipitation which does not reach the ground; in this case the base of the cloud is always diffuse and looks "wet" on account of the general trailing precipitation, VIRGA, so that it is difficult to determine the limit of its lower surface.'

9. CUMULUS (CU) are isolated clouds developing *upward* with generally FAIR type of weather,

THICK CLOUDS WITH VERTICAL DEVELOPMENT; THE UPPER SURFACE IS DOME SHAPED AND EXHIBITS ROUNDED PROTUBERANCES, WHILE THE BASE IS NEARLY HORIZONTAL.

'When the cloud is opposite the sun the surfaces normal to the observer are brighter than the edges of the protuberances. When the light comes from the side, the clouds exhibit strong contrasts of light and shade; against the sun, on the other hand, they look dark with a bright edge. True cumulus is definitely limited above and below; its surface often appears hard and clear cut. But one may also observe a cloud resembling ragged cumulus in which the different parts show constant change. This is designated FRACTOCUMULUS.'

10. CUMULONIMBUS (CB), associated with SHOWERY weather, are

HEAVY MASSES OF CLOUD, WITH GREAT VERTICAL DEVELOPMENT, WHOSE CUMULIFORM SUMMITS RISE IN THE FORM OF MOUNTAINS OR TOWERS, THE UPPER PARTS HAVING A FIBROUS TEXTURE AND OFTEN SPREADING OUT IN THE SHAPE OF AN ANVIL.

'The base resembles nimbostratus, and one generally notices *virga*. This base has often a layer of very low ragged clouds below it (fractostratus, fractocumulus). Cumulonimbus clouds generally produce showers of rain or snow and sometimes of hail or soft hail, and often thunderstorms as well. If the whole of the cloud cannot be seen the fall of a shower is enough to characterise the cloud as cumulonimbus.'

The tables on pp. 14–20 give further details.

HIGH CLOUD TYPES IN DETAIL

C_H no.	Official definition	Official description	Meaning in this book
0	No Ci, Cc or Cs		No high clouds
1	Filaments or strands of cirrus, scattered and not increasing, often 'mares' tails' (Ci filosus)	Wisps of cloud at a very high level; they may be scattered over a large part of the sky but the amount does not increase noticeably either in time or in any particular direction. The clouds do not collect into sheets and bands, and there is no tendency for the elements to fuse together into masses of cirrostratus	FINE CI, NOT INCREASING, SPARSE, merely showing the very high and cold upper air remaining locally just saturated by its very small quota of moisture, quite innocuous
2	Dense cirrus in patches or twisted sheaves usually not increasing, possibly but not certainly the remains of the upper part of cumulonimbus (Ci densus)	Cirrus of this type is more 'woolly' in appearance than C_HI and is possibly, but not certainly, the debris of the upper part of cumulonimbus	FINE CI NOT INCREASING: ABUNDANT, BUT NOT A CONTINUOUS LAYER, still innocuous as long as it is not increasing, though it may be a sign of upper air disturbance just past
3	Cirrus, often anvil-shaped, either the remains of the upper portions of cumulonimbus or part of a distant cumulonimbus the rest of which is not visible (Ci nothus). If there is doubt as to the cumulonimbus origin or association, code figure 2 is used	Cirrus, often anvil-shaped, usually dense, which is known to be either the remains of the upper part of a disintegrated cumulonimbus or part of a distant cumulonimbus the rest of which is not visible at the time of observation	'ANVIL', 'FALSE' OR DENSE CI, associated with, if not actually attached to, Cb tops
4	Cirrus (often hook shaped) gradually spreading over the sky and usually thickening as a whole	This type of cirrus, which is in the form of streaks ending in a little upturned hook or in a small tuft, increases in amount both in time and in a certain direction. In this direction it reaches to the horizon, where there is a tendency for the cloud elements to fuse together, but the clouds do not pass into cirrostratus	FINE CI INCREASING, USUALLY IN TUFTS, first sign of disturbance approaching, sometimes with no further development, but often associated with type 5 described below
5	Cirrus and cirrostratus, often in bands converging toward the horizon, or cirrostratus alone; in either case gradually spreading over the sky and usually thickening as a whole, but the continuous layer not reaching 45° altitude	Sheet of fibrous cirrus partly uniting into cirrostratus, especially towards the horizon in the direction where the cirrus strands tend to fuse together; the cirrus is often in a herring-bone formation or in great bands converging more or less to a point on the horizon. In this class is also included a sheet of cirrostratus which does not cover the sky and is below 45° altitude	CI OR CS INCREASING, STILL BELOW 45° ALTITUDE above the horizon, OFTEN IN POLAR BANDS, i.e. in parallel bands right across the sky

6

Cirrus and cirrostratus, often in bands converging toward the horizon, or cirrostratus alone; in either case gradually spreading over the sky and usually thickening as a whole, and the continuous layer exceeding 45° altitude

The definition of this type is the same as the previous one, with the exception that the cloud reaches more than 45° above the horizon. (Note that altitudes if not measured instrumentally are deceptive; it is common to overestimate a point in the sky. A point at 30° altitude will appear to be about 45° altitude)

CI OR CS INCREASING, REACHING ABOVE 45° ALTITUDE

7

Cirrostratus covering the whole sky

This is either
(a) a thin uniform nebulous veil, sometimes hardly visible, sometimes relatively dense, always without definite detail, but producing halo phenomena round the sun and moon; or
(b) a white fibrous sheet, with more or less clearly defined fibres, often like a sheet of fibrous cirrus from which indeed it may be derived

VEIL OF CS COVERING WHOLE SKY (halo round sun or moon), such as may merge into As and thus into rain or snow-fall when some air mass-lifting disturbance approaches. Hence the idea of a halo often portending bad weather

8

Cirrostratus not increasing and not covering the whole sky; cirrus and cirrocumulus may be present

This is a case of veil or sheet cirrostratus reaching the horizon in one direction but leaving a segment of blue sky in the other. This segment of blue sky does not grow smaller, otherwise it would be reported as $C_{H}5$ or $C_{H}6$. Generally the edge of the sheet is clear-cut and does not tail off into scattered cirrus

CS NOT INCREASING AND NOT COVERING WHOLE SKY, usually a sign of some upper air mass lifting but disturbance past

9

Cirrocumulus alone or cirrocumulus with some cirrus or cirrostratus, but the cirrocumulus being the main cirriform cloud present. Cirrocumulus may be present in $C_{H}1$ to $C_{H}8$

Cirrocumulus is a wavy type of 'mackerel' sky with a delicate fine structure. Cirrocumulus is not to be confused with small altocumulus. There must be either evident connexion with cirrus or cirrostratus, or the cloud observed must result from a change in cirrus or cirrostratus

CC PREDOMINATING, AND A LITTLE CI

MEDIUM CLOUD TYPES IN DETAIL

C_M no.	Official definition	Official description	Meaning in this book	Height of the order of
o	No Ac, As or Ns	—	No medium clouds	—
1	Thin altostratus (semi-transparent) through which the sun or moon would be seen dimly as through ground glass (As translucidus)	This is a darkish veil usually covering the whole sky, though not always. It looks rather like a thinly fogged photographic plate. The sun or moon appears as though shining through ground glass and does not cast a shadow. Halo phenomena are not seen in altostratus. A sheet of this cloud resembles thick cirrostratus (see C_{H7}) from which it is often derived	THIN ALTOSTRATUS	5 km.
2	Thick altostratus or nimbostratus. Through portions of the sheet the position of the sun or moon may be indicated by a light patch (As opacus or nimbostratus)	The sun and moon are generally hidden or are indicated only by the lighter colour of one part of the cloud. Typical thick altostratus can be formed either by a thickening of thin altostratus or by the fusing together of the cloudlets in a sheet of altocumulus	THICK ALTOSTRATUS (sun or moon invisible) or Ns, usually raining or snowing from high up, and formed by damp air mass rising gradually	Top above 5 km., base below 5 km., often below 3 km.
3	Thin (semi-transparent) altocumulus, cloud elements not changing much, at a single level	Altocumulus often looks like sheep's fleeces. This type generally forms a single layer; it is fairly regular, and of uniform thickness, the cloudlets always being separated by clear spaces or lighter gaps; the cloudlets are neither very large nor very dark. This layer is generally fairly persistent, it does not change or disappear quickly. Cases of altocumulus which are so dense that the waves do not show lighter parts should be reported as C_{M7}	SINGLE LAYER OF Ac OR HIGH Sc, thin and harmless, with a few breaks or clear patches	2–3 km.

No.				
4	Thin (semi-transparent) altocumulus (often in patches almond- or fish-shaped), cloud elements continually changing and/or occurring at more than one level	The cloudlets may be as small as cirrocumulus, but lenticular altocumulus shows delicate colouring (irisation). Where this is so, the clouds are often scattered over the sky quite irregularly and may be at different levels. Though individually they may be changing, the amount of cloud over the whole sky generally remains about the same	AC IN ISOLATED BANDS, IN-DIVIDUALLY DECREASING, OFTEN LENTICULAR (thickest in the middle, and tapering to the edge, like a lens), apt to appear where upper air on the verge of saturation is flowing *almost* horizontally but with gradual slight ups and downs, as in weather which is unsettled though not actually bad	4–6 km.
5	Thin (semi-transparent) altocumulus in bands or in a layer gradually spreading over the sky and usually thickening as a whole; it may become partly opaque or double-layered	Either the bands are great elongated masses, sometimes appearing rather dark, often of a roughly lenticular shape, or the ordinary altocumulus waves are crossed by blue lanes, so that they appear like bands (with the waves across the bands). An essential feature of this type is that the sky becomes more and more covered	AC IN BANDS, INCREASING, tends to form where upper air on the verge of saturation is rubbed over and thereby kept stirred (in a shallow layer) by a faster flowing air-stream, of which it is thus a characteristic sign	4–6 km.
6	Altocumulus formed by the spreading out of cumulus (Ac cumulogenitus)	Cumulus clouds of sufficiently great vertical development may undergo an extension of their summits while their bases may gradually melt away. The process is similar to that of C_{L4} but at a higher level. The cloud which looks anvil-shaped must not be confused with C_{L9}	AC FORMED FROM CU SPREADING OUT, i.e. from underneath, rather than by the upper air currents themselves	3–4 km.

MEDIUM CLOUD TYPES IN DETAIL (*cont.*)

C$_M$ no.	Official definition	Official description	Meaning in this book	Height of the order of
7	Any of the following cases: (*a*) double-layered altocumulus usually opaque in parts, not increasing, (*b*) a thick (opaque) layer of altocumulus, not increasing, (*c*) altostratus and altocumulus both present at the same or different levels		AC ASSOCIATED WITH AS, or AS WITH PARTS RESEMBLING AC, which, like ordinary As, is a sign of upper air mass rising and apt to lead to rain or snow	5 km.
8	Altocumulus in the form of cumulus-shaped tufts or altocumulus with turrets (Ac castellatus or Ac floccus)	Altocumulus castellatus is composed of small cumuliform masses with more or less vertical development, either detached or forming a band. Altocumulus floccus cloudlets are of ragged appearance without definite shadows and with the rounded parts slightly domed. There are often pronounced trails (*virga*) of cirriform appearance	AC CASTELLATUS (OR AC IN RAGGED FRAGMENTS). Like Cu, growing upward more than horizontally, sometimes therefore in the form of little columns, like battlements on a castle (whence the name castellatus), only at medium instead of low cloud levels, this form is a well-known sign of thundery tendency in the upper air, appearing before a thunderstorm rather than after it	3–4 km.
9	Altocumulus of a chaotic sky, generally at different levels; dense cirrus in patches usually also present	The sky has a disordered, heavy and stagnant appearance. It is very complex with patches of medium cloud more or less fragmentary, superposed, often badly defined and giving all the transitional forms between low altocumulus and the fibrous veil	AC IN SEVERAL LAYERS, GENERALLY ASSOCIATED WITH FIBROUS VEILS AND A CHAOTIC APPEARANCE OF THE SKY, especially after a thunderstorm which has largely developed in a mass of saturated air high up, of which it might then be described as the outlying wreckage	4–8 km.

18

LOW CLOUD TYPES IN DETAIL

C_L no.	Official definition	Official description	Meaning in this book	Height in km. of the order of	
				Base	Top
o	No Sc, St, Cu or Cb	—	No low clouds	—	—
1	Cumulus with little vertical development and seemingly flattened (Cu humilis)	The clouds look rather like cauliflowers. The bases tend to be flat, and to be at a uniform level. They are scattered and have a flat and deflated appearance, even when convection is greatest in the early afternoon. Their horizontal extension is greater than the vertical	FAIR WEATHER Cu, small or broken, formed only in air over warmer land or sea, and unable to rise very far	1	1-2
2	Cumulus of considerable development, generally towering, with or without other cumulus or stratocumulus, bases all at the same level (Cu congestus)	The difference between these and the fair weather cumulus is that the tops of the clouds instead of remaining rounded (and apparently quiescent) begin to bulge upwards and 'rising heads' appear	LARGE Cu, with rounded tops, marking rising air currents, with clear (descending) air in between	1-2	2-4
3	Cumulonimbus with tops lacking clear-cut outlines but distinctly not cirriform or anvil-shaped, with or without cumulus, stratocumulus or stratus (Cb calvus)	Distinguished from type 2 by the fact that the tops are beginning to acquire a fibrous appearance and by showers falling from the base	Cb, shower or thunder cloud	1-2	Over 4
4	Stratocumulus formed by the spreading out of cumulus; cumulus also often present (Sc cumulogenitus or Sc vesperalis)	This cloud is formed in two ways: (a) during the day when there is a stable layer or an inversion which the convective cumulus clouds reach and cannot penetrate; (b) in the evening when convection weakens, with or without an inversion above the cumulus. It is most common in the evening	Sc FORMED BY Cu SPREADING OUT, when unable to grow any further upward, but not yet evaporated away. An innocuous type	1-2	2-3

LOW CLOUD TYPES IN DETAIL (*cont.*)

C_L no.	Official definition	Official description	Meaning in this book	Height in km. of the order of	
				Base	Top
5	Stratocumulus not formed by the spreading out of cumulus	The individual cloud masses may be detached and more or less lenticular in shape or close together in a continuous (or nearly continuous) layer. Stratocumulus is often a dark cloud particularly in winter, but it may be fairly light—usually when it is at a fairly high level	STRATOCUMULUS (Sc) OR STRATUS (St) layer, never very thick, but common in all damp air which is cooled from underneath or on hills. Unlike Cu it is 'all or nothing'—being NEARLY ALWAYS EITHER SPARSE OR WIDESPREAD, and can precipitate only drizzle (or its ice-crystal equivalent)	Sc base and top 1–2 but STRATUS base below ½ km. and top normally below 1 km.	
6	Stratus or fractostratus or both, but not fractostratus of bad weather	Stratus is a uniform layer of cloud, resembling fog, but not resting on the ground. When this very low layer is broken up into irregular shreds it is designated fractostratus	(Not used in this book, being covered by the old code figure 5 whilst the cloud type covered by the old No. 6 is now called No. 7, described below)	—	—
7	Fractostratus and/or fractocumulus of bad weather ('scud') usually under nimbostratus. By 'bad weather' is meant the conditions prevailing before, during or after precipitation	These low clouds, collectively known as FRACTONIMBUS, often show up very dark against the relatively lighter background of altostratus or nimbostratus	Ragged low clouds of bad weather (formerly called NIMBUS), unable to precipitate by themselves but normally formed in rain or snow that falls out of upper clouds	Base generally below ½ km., top below 1 km.	
8	Cumulus and stratocumulus other than those formed by the spreading out of cumulus, with bases at different levels		Cu WITH A Sc LAYER ABOVE IT, normally formed independently from different sources	Cu base 1 km., Sc base 2 km. or higher	
9	Cumulonimbus having a clearly fibrous (cirriform) top, often anvil-shaped, with or without cumulus, stratocumulus, stratus or 'scud'	Ragged low clouds of bad weather are often present. The anvil-shaped mass of cirriform cloud may be hidden from a near-by observer by lower parts of the cloud mass. It is important therefore to keep a close watch on the sky to ensure accurate differentiation between C_L3 and C_L9	LARGE Cu OR Cb WITH RAGGED LOW CLOUDS, usually associated with extensive showers or storms	—	—

Column 9 gives the VISIBILITY over land, estimated from landmarks, thus:

Code number	0	1	2	3	4
Kilometres	Less than 0·05	0·05–0·2	0·2–0·5	0·5–1·0	1–2
Visibility type	Dense fog	Thick fog	Moderate fog		Mist

Code number	5	6	7	8	9
Kilometres	2–4	4–10	10–20	20–50	More than 50
Visibility type	Slight mist or haze		Good vis.	Very good vis.	Excellent vis.

The new international code would add a '9' in front of our single code-figure, or else report more exactly (if between 0·2 and 16 km.) in furlongs or fifths of a kilometre. Anyway it gives *two* figures instead of our *one*.

Column 10 gives mean WIND direction (as shown by the nearest well-exposed wind vane or smoke drift or, next best, by clouds below about 1500 ft. or $\frac{1}{2}$ km.).

Column 11 gives wind force on the Beaufort Scale thus:

No.	Knots	Description	Specification for use on land
0	<1	Calm	Smoke rises vertically
1	1–3	Light air	Direction of wind just shown by smoke drift
2	4–6	Slight breeze	Wind felt on face; leaves rustle
3	7–10	Gentle breeze	Leaves and small twigs in constant motion; wind extends light flag
4	11–16	Moderate breeze	Raises dust and loose paper; small branches are moved
5	17–21	Fresh breeze	Small trees in leaf begin to sway; crested wavelets appear on inland waters
6	22–27	Strong breeze	Large branches in motion; whistling heard in telegraph wires
7	28–33	Moderate gale	Whole trees in motion; inconvenience felt when walking against wind
8	34–40	Fresh gale	Breaks twigs off trees, and generally impedes progress

Here, too, the new international code would replace our single figure by a double figure (the wind speed in knots at effective height of 10 m. over the ground).

Columns 12 and 13 give the shade TEMPERATURE in degrees F. and C. The thermometer at station No. 18 (Dulwich) had standard exposure in a Stevenson screen, while the thermometer at station No. 1 was against the north wall of a house. Temperatures have been recorded to the nearest tenth of a degree Fahrenheit and then converted (for purposes of the commentary, where for simplicity we prefer the metric units which are more nearly universal in science) into Centigrade to the nearest degree.

CHARTS are introduced later in Chapter VIII.

III

WEATHER TYPES

1. FIRST OBSERVATIONS

Place no.	Date 1936	Time G.M.T.	Weather letters	Cloud				Vis. V.	Wind		Temp.	
				T.	L.	M.	H.		D.	F.	F.	C.
01	15.8	0730	b	1	0	0	9	6	SW'S	1	66·6	19
01	15.8	1200	bc	3	1	0	0	6	SW'S	2		
01	15.8	1315	b	1	1	0	9	7	SW'S	2	76·0	24
02	15.8	1530	bc	4	1	0	9	7	SW'S	3		
01	15.8	1700	b	1	0	6	9	7	SW'S	2	75·5	24

First observations recorded are on **15 August 1936**, a simple day of fair weather type showing normal features:

Morning (0730)

Weather b=fine: clear sky
Total cloud amount 1=only one-eighth of the sky
Low cloud type o=no low cloud
Visibility 6=slight mist (4–10 km.)
Wind direction SW'S=south-west by south
Wind force 1=light air
Temperature 19=19° C., just cool (by summer standards)

Afternoon

Weather bc=partly cloudy
Low cloud type 1=small cumulus, increasing to
Total amount 4=four-eighths of the sky when
Temperature is at its highest
Wind force 2–3 is more than it was this morning, and
Visibility is now 7, which is good.

Evening (1700)

Weather again is b=clear sky; total cloud amount has decreased; again there is no low cloud, and both wind force and temperature have just started to fall.

Being south-westerly, the wind comes from the sea. Hence its moisture for early morning mist, even with temperature nearly 20° C.

The upper clouds are innocuous.

2. FRONTS

Place no.	Date 1936	Time G.M.T.	Weather letters	Cloud				Vis. V.	Wind D.	F.	Temp. F.	C.
				T.	L.	M.	H.					
01	16.8	0500	b	1	0	0	9	6	SW'W	3	61·5	16
01	16.8	0600	b	0	0	0	0	6	SW'W	3		
01	16.8	1200	b	1	1	0	0	7	SW'W	4	78·0	26
01	16.8	1500	bc	4	1	0	0	7	SW'W	4	79·8	27
01	16.8	2000	b	1	4	0	0	7	WSW	3	73·3	23
01	17.8	0730	c	6	8	0	0	6	W'S	3	74·0	23
01	17.8	0745	b	1	0	0	1	6	W'S	3		
03	17.8	1230	bc	4	1	0	0	7	W	3		
01	17.8	1930	b	2	0	0	9	7	W	4	69·7	21
01	17.8	2100	c	7	5	–	–	7	W	4	68·0	20
01	18.8	0600	o	8	5	–	–	6	W	3	66·2	19
01	18.8	1030	o	8	5	–	–	6	W	3		
01	18.8	1600	bc	2	2	0	0	7	W'N	4	73·2	23
01	18.8	1900	b	1	0	0	5	7	WNW	4	69·0	21
01	18.8	2100	c	7	0	1	–	7	WNW	4		
01	19.8	0900	o	8	0	2	–	7	W'S	5	63·0	17
01	19.8	0945	o	8	0	2	–	7	SW	5		
01	19.8	1000	o	8	0	7	–	7	SW	5		
01	19.8	1330	o	8	0	7	–	8	SW	5		
01	19.8	1500	rr	8	0	2	–	7	SSW	6		
01	19.8	1700	rr	8	7	–	–	7	SSW	6		
01	19.8	1800	r_0r_0	8	5	–	–	7	SSW	6	60·4	16
01	19.8	1900	d	8	5	–	–	7	SSW	6	61·0	16
01	20.8	0800	c	7	5	–	–	7	WNW	3	65·8	19
01	20.8	1000	c	7	2	–	–	7	WNW	3		
01	20.8	1300	c	6	3	–	6	7	WNW	3	74·0	23
01	20.8	1545	bc	4	2	0	8	7	WNW	4	73·6	23
01	20.8	1645	b	2	0	7	9	7	W	4	73·0	23
01	20.8	1800	b	2	0	0	1	7	WSW	4	70·0	21
[a] 01	20.8	1900	b	1	5	0	1	7	WSW	4	68·8	20

[a] Cirrus moving from WNW.

16 August is like 15 August.

17 August is similar but a bit cooler, with low clouds also at the beginning and end of the day.

18 August is similar but with more stratus clouds in the morning followed by larger cumulus clouds in the afternoon, suggesting cooler air aloft.

The uplift of air over hills helps clouds to form, but uplift on fronts is often much more important, as warm air slides up over cold. The clouds are of characteristic types, high and medium clouds often unbroken, extensive and thick, in which case the frozen parts give snow which by melting forms rain, while the rest gives drizzle. Precipitation is generally of continuous type, not showers.

To see what a front does, we ask what it is. The polar front is the edge of an ocean of polar air. The arctic, for instance, is always submerged, whilst the tropics, 'higher and drier' so to speak, remain always warm. Between them the waves wash over the world, the ocean's edge advancing into the tropics as a kind of tide comes in. How does it go out again? Just as a wet seashore becomes dry like the unwetted land, so the flood of cold air becomes warm like the uncooled tropics. Just as the seashore stays wet all the time underneath but turns into dry land on top, so the cold air stays dry on top but turns into tropical wet air below. Being dense, too, it brings rising pressure and therefore subsidence and warmth aloft.

As the state of the seashore just depends on the water's edge, which is the boundary line upon the land surface, so the weather depends upon the positions of the atmosphere's fronts or air-mass boundary lines on the actual surface of the earth or sea. The 'positions of fronts' of which we speak are these. Naturally they shift a long way for only slight rises or falls of the ocean of air. In northern latitudes their northernmost points (the crests of the waves as shown on a map, though really the troughs of the waves of cold air) are low-pressure centres round which blow the winds.

Ahead of each such wave we have warm air overrunning colder air which lies in the form of a wedge underneath it. Thus laden with cloud it climbs a slope. Highest and farthest ahead are increasing cirrus and cirrostratus clouds. A halo appears round the sun or moon which then looks more and more watery until completely obscured. Rain or snow begins to fall from the deepening grey altostratus or nimbostratus cloud which goes on until the front at ground level has passed and so brought in behind it the new warm air mass.

This part of the wave, then, with cold air ahead but warm air behind it, is called the depression's WARM FRONT. At the other part of the wave, the COLD FRONT, everything is reversed, the cold air chasing and tending to under-cut the warm. In the WARM SECTOR between them over temperate seas there is apt to be FOG which then becomes stirred up inland into much low cloud of the STRATUS type.

19 August shows this warm-front weather with

(a) from the previous evening, C_H type $5 = Ci$ and Cs clouds which are only low down in the sky at first but increasing to C_M type 1 (altostratus) which covers the sky, becoming

(b) this morning, overcast and then rainy with nimbostratus and ragged low clouds until evening when rain decreases, cloud-type changes, and temperature rises slightly. The warmer air mass has arrived.

Next after this comes more polar air with only broken cumulus clouds. In fact as it overtakes the warm air which in turn is overtaking the old polar air you see that the *cold* front which is the warm sector's *rear* edge must overtake the warm front. They then coalesce as a single front, an OCCLUSION, where the old warm air is lifted up off the ground altogether.

The air on the ground is then all polar, but not quite the same ahead as the air behind the occlusion, where it is fresher from its cold source. Yet the old air ahead may have come from a source so much colder and/or drier as to leave it still effectively colder than the air behind the occlusion. The occlusion is then in effect a warm front, so we call it a WARM OCCLUSION.

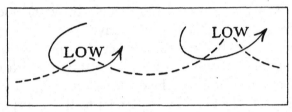

Fig. 4. Successive wave cyclones.

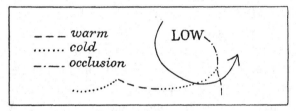

Fig. 5. Occlusion.

If, on the other hand, the air behind the occlusion is colder, then we have in effect a cold front. That is a COLD OCCLUSION. The coalescence of the old warm and cold fronts takes place from the depression's centre outwards, so that an open though ever-narrowing warm sector will remain somewhere on the tropical side of the occlusion. Its cold front trailing out somewhere to west is thus linked on to the next developing wave.

In the RIDGE of higher pressure between the old and the new depression thus formed, subsidence damps out convection, or rather more literally *dries* it out, tending to disperse the large cumulus (C_L type 2) and even cumulonimbus (C_L type 3 or 9) which naturally develop at first in this fresh polar air as it comes over warmer lands and seas. So any showers which may come in between the bright intervals characteristic of the air behind the cold front or occlusion will die out in the ridge between two depressions until, at length, all the tops of the cumulus clouds are cut off by the warm upper air advancing again ahead of the next warm front. And so the chain of depressions goes on, no two links quite the same. In the end it is broken by the growth of one ridge into a big high-pressure area or anticyclone. Which of the ridges will do so is hard to predict, but when it is coming it makes the next wave appear to hang back. The next depression thus never arrives but is steered elsewhere. An effect like this is described as a BLOCKING action.

20 August shows polar-air weather becoming fair (bc) or fine (b), with decreasing amount of low cloud (types 2 and 3 becoming 0), a veer of direction (from south-south-west to west-north-west) and decrease of force of wind (from 6 to 3–4) since yesterday.

A front, you see, is simply the boundary between *any* two different masses of air, whether sudden or only gradual in transition. With the effectively warmer air always sloping up over the colder air mass, the front is bound to share their motion. If the air-flow is parallel to the front, then obviously the front stays still, QUASI-STATIONARY, well defined by contrast of air mass properties such as weather type, yet inactive with no ascent of the warmer air over the colder. Medium cloud types 4 or 5, for instance, tend to appear rather than types 1 or 2. Innocuous high cloud types 1, 2, 8 or 9 appear rather than 4, 5, 6 or 7. But if the warm air overtakes the colder air at an angle, even without any 'total' change taking place in either mass of air as it goes along, then the bodily transport or advection of the new air is bound to appear as a change of air at any one place underneath. Whether sudden or only gradual, this is the *local change* caused by the movement of a front. If the advance is of warm against cold, not cold against warm, then the front is regarded as warm, a warm front. If, on the other hand, the advance is of cold against warmer air, then the same front must be described as a cold front. The warm air, you see, lies all the time on top of the wedge of cold, but a slight alteration of their *relative flow* will make all the difference between warm-front and cold-front weather. On a warm front generally the warm wet air is gradually forced ever higher up the slope as it flows along. Hence extensive sheets of upper cloud. Above a cold front, on the other hand, it may be advancing too fast for the cold air and therefore slides ever lower down the slope, subsiding and so getting warmer. Hence the upper clouds' tendency to clear.

Thus the same transition zone between different air masses can produce quite different weather types purely according to the angle at which the

26

air masses flow against it—whether at ZERO angle (quasi-stationary front), at an angle which points INTO THE COLDER AIR (warm front), or at an angle which points INTO WARMER AIR. Not only upon direction but also on *speed* of air-flow depends the relative flow of the two air masses and thus the activity of the front, which is therefore least at a high-pressure centre where the air is almost calm. High pressure also causes general outflow over the ground. To make up for this horizontal outflow the upper air flows in downwards and so counteracts upflow on a front, thus tending to clear the bad weather on the spot instead of waiting for it to shift. Low pressure, on the other hand, makes things worse. Minor cold fronts, for instance, between rather similar polar-air masses, are apt to cause heavy showers round a depression whilst innocuous round an anticyclone.

Not only on relative air-flow, of course, but upon how sudden or gradual is the frontal transition does the weather depend. Born as two air masses meet, a front will intensify if the air-flow pattern is such as to squeeze the air masses together and make the transition zone narrow so that all 'contour' lines which mark out the different air-mass properties are closely crowded. A change of flow pattern which draws them apart again will weaken the front's activity. There is indeed more in the beautiful patterns of air-flow than meets the eye.

3. WIND

21 August is like 20 August, and we turn to the subject of wind.

How does water flow over land? It depends on the lie of the land, that is, the distribution of HEIGHT of the land, which is mapped out by CONTOURS.

Likewise at any one height the distribution of PRESSURE, mapped out by ISOBARS, determines the air-flow or wind. We call the distribution of pressure a 'high', a 'low' or depression, a ridge, a trough or a col, according as the isobars are arranged like the contours of a hill, a valley or depression, a ridge, a trough or a col. It is again, you see, a matter of pattern.

Suppose there is for a while no change in the pattern. Then apart from friction we recognise three forces:

PRESSURE GRADIENT FORCE. This force, acting from high to low pressure, is measured by the pressure gradient or isobar spacing, just as the spacing of contours measures the gradient or slope of the land.

GEOSTROPHIC FORCE. To the right of the wind direction in northern latitudes, or to the left in southern, a so-called geostrophic force is recognised, strong in proportion to the wind speed itself.

CENTRIFUGAL FORCE. Centrifugal or centripetal force is the familiar force recognised whenever the wind—or anything else—moves in a curve instead of in a straight line.

If the wind flows straight, we can disregard centrifugal or centripetal force, and therefore say that when all is steady the first two forces alone must balance each other. As the geostrophic force depends on the wind while the pressure-gradient force depends on the pressure gradient, the steady wind that blows when they balance can only depend on the pressure gradient which the isobars show. That is the GEOSTROPHIC WIND.

Curvature, bringing in centrifugal or centripetal force, makes the true wind lighter than geostrophic around a low-pressure centre, but stronger than geostrophic around a high. The fictitious but often almost true wind deduced by assuming the balance of all three forces is called the GRADIENT WIND.

	Place no.	Date 1936	Time G.M.T.	Weather letters	Cloud				Vis. V.	Wind		Temp.	
					T.	L.	M.	H.		D.	F.	F.	C.
	01	21.8	0615			7	0	9	6	W	3	61·5	16
	01	21.8	0700	bc	3	4	0	9	6	W	3		
	01	21.8	1000	c	7	0	3	–	6	W'N	3		
	01	21.8	1215	c	7	3	–	–	6	WNW	2	69·9	21
	01	21.8	1300	u	8	3	–	–	6	WNW	3	69·9	21
	01	21.8	1500	bc	5	3	8	0	7	WNW	2	71·0	22
	01	21.8	1515	c jp	6	3	0	9	7	WNW	3		
	01	21.8	1600	bc	5	3	0	0	7	WNW	4	69·9	21
	01	21.8	1830	c	7	8	–	–	7	WNW	3	68·1	20
	01	21.8	2030	bc	4	0	7	–	6	WNW	4		
	01	22.8	0715	b m$_0$	0	0	0	0	5	WNW	2	60·8	16
	01	22.8	0800	b m$_0$	0	0	0	0	5	WNW	1	59·1	15
	01	22.8	0830	b m$_0$	0	0	0	0	5	WNW	2	63·1	17
	01	22.8	1000	b	2	1	0	0	7	WNW	4	66·3	19
	02	22.8	1100	bc	6	3	0	0	7	WNW	3		
	01	22.8	1300	bc	5	0	6	2	7	WNW	2		
[a]	02	22.8	1330	bc	5	0	–	2	7	WNW	2		
	02	22.8	1415	u	7	3	–	–	7	WNW	2		
	01	22.8	1550	bc	6	4	–	–	7	WNW	3	68·2	20
	01	22.8	1830	b	2	0	0	9	7	WNW	3		
	01	23.8	0400	b	1	5	0	0	6	Calm			
	01	23.8	0530	b m$_0$	1	0	0	1	5	W	1	58·4	15
	01	23.8	0830	b m$_0$	0	0	0	0	5	W	1	61·5	16
	04	23.8	0910	b	0	0	0	0	6	W	2		
	01	23.8	1400	b	0	0	0	0	8	W	4	72·8	23
	01	23.8	1600	b	0	0	0	0	8	W	4	74·0	23
	01	23.8	1930	b	0	0	0	0	6	WSW	3		
	01	24.8	0500	b	0	0	0	0	6	SW	3	60·1	16
	01	24.8	1530	b	1	1	0	1	7	W'N	4	79·0	26
	01	24.8	2140	b	0	0	0	0	6	W'N	3	71·9	22

[a] Cirrus moving from N.

Head-on FRICTION, as produced by obstacles or even mere roughnesses of the earth's surface (whether land or water), naturally slows the wind down. What is not quite so obvious (unless you remember that the deflecting effect of the earth's rotation goes on all the time), is that when thus retarded the wind appears deflected at an angle across the isobars towards lower pressure—most noticeably, of course, at the earth's surface, less so aloft, until at a height of 1 km., or even at 2000 ft., the drag effect is often negligible.

The true 2000 ft. wind, then, is nearly enough the same as the gradient wind at that height. But in turn the gradient (of pressure) at that height is nearly enough the same as the gradient at sea-level. So the 2000 ft. wind is determined by the isobars of pressure reduced to mean sea-level, such as are drawn on ordinary weather maps.

What does this pressure mean? It represents the weight of all the air above sea-level, allowing for thickness of land as well. The land above sea-level is imagined to be replaced by air. The 10,000 ft. or 3 km. pressure likewise represents the weight of all the air above 3 km. So the pressure *difference* must represent the weight of the bottom 3 km. of the atmosphere. But that depends on its average density, which depends in turn on its temperature. So it is on the mean *temperature* distribution that depends the difference between low-level and high-level wind.

This (vector) difference between low-level and high-level geostrophic winds is known as the THERMAL wind. As it blows round the high- or low-temperature centres in just the same way as the total geostrophic wind blows round the high- or low-pressure centres, we have the following well-known rules:

1. Up to about 1 km., wind strengthens and veers with height;
2. Above about 1 km., wind veering with height brings warmer air, wind backing with height brings colder air; wind strengthening with height means that temperature is high where pressure is high, whereas a wind that decreases with height means temperature high where pressure is low, or cold where pressure is high.

We supposed no change in the pressure pattern. RISING OR FALLING PRESSURE, which changes the pattern, has an effect of pushing the air across the isobars towards the region of most rapidly falling (or least rapidly rising) pressure. Combined with the drift across the isobars due to friction it causes outflow or DIVERGENCE from centres of HIGH or RISING pressure, with CONVERGENCE into LOW or FALLING pressure. To make up for excessive horizontal convergence the air must rise, whereas to make up for excessive divergence it sinks; and you know how much the air's rising or sinking affects the weather. In a frontal wave depression or cyclone you know that the different air masses have different flow. The different flow of the different air masses implies a different run of the isobars, which therefore kink at the fronts. So the cross-isobar wind components (due to friction or unevenly changing pressure) converge on the fronts, thus helping to make the air rise and condense its moisture into bad weather.

Ahead of a warm front the warm air is, of course, deepening over any fixed place, whereas behind the front it can deepen no further. So there must be an abrupt change of rate of fall of total pressure, as is implied by the kink in the isobars passing by. Behind a cold front, too, cold air comes in an ever-deepening wedge, with the same effect. The kink in the isobars may be sharp or blunt, but is always pointing towards the high pressure.

22 August is like 21 August except for slight mist at first without any clouds. This makes it almost normal fair-weather type, as the cumulus clouds reach maximum development in the middle of the afternoon and then spread out before clearing away.

The cirrus and its direction of motion suggest a new warm front lying nearly north-south and not readily coming in: probably a ridge of high pressure developing to hold it up. For this the broad argument is that the wind veers a lot with height (from west-north-west at the surface, probably north-west gradient wind at 1 km., to northerly at high cloud levels). More exactly, what is the thermal wind? It is the vector difference between a light wind coming from west of north, and a strong upper wind from the north. So it must itself be quite a strong wind from just east of north. But thermal wind blows with colder air on its left than on its right. So to-day it is colder to east than to west. In fact we should say there was a warm sector to west. So there must be a north-south warm front.

23 August is a day of fine weather type fairly typical of high pressure in summer, showing normal features as on 15–16 August but with less cloud and slightly smaller temperature range. Notice how the visibility is apt to fluctuate during the first two or three hours of daylight as the new day's warmth just stirs up the damp morning air.

24 August is similar. Weather is fine (b); cloud is only a trace of small cumulus at mid-afternoon; visibility is good except for slight mist early morning and evening, while temperature ranges through 10° C.

30

4. FINE WEATHER TYPE

Place no.	Date 1936	Time G.M.T.	Weather letters	Cloud				Vis. V.	Wind		Temp.	
				T.	L.	M.	H.		D.	F.	F.	C.
01	25.8	0600	c m	7	0	3	–	4	WNW	1	65·8	19
05	25.8	1030	c m$_0$	7	3	–	–	5	NW'W	2		
06	25.8	1200	bc	4	1	0	0	6	NW	2		
07	25.8	1230	b	2	1	0	0	6	NW	2		
01	25.8	1800	b	0	0	0	0	6	NW	2		
01	26.8	0700	c	7	5	2	–	7	ENE	5		
01	26.8	0915	c	7	5	–	–	7	NE	6		
01	26.8	1400	b	1	1	0	0	8	NE	6	68·9	20
01	26.8	1630	b	0	0	0	0	8	NE	5	65·4	19
01	27.8	0600	b m$_0$	0	0	0	0	5	E	2		
01	27.8	0745	b	2	8	0	0	6	E	3	65·0	18
01	27.8	0840	c	7	8	0	0	7	E	4	65·5	19
01	27.8	1000	bc	4	5	0	0	7	E	4	65·5	19
08	27.8	1400	b	0	0	0	0	6	E	4		
08	27.8	1800	b	0	0	0	0	6	E	2		
08	28.8	0800	c	7	8	1	–	6	E	2		
08	28.8	0900	c	6	8	–	–	6	E	3		
08	28.8	1000	b	0	0	0	0	7	E	3		
08	28.8	2100	b m$_0$	0	0	0	0	5	E	1		
08	29.8	0800	b	0	0	0	0	8	Calm			
08	29.8	1900	b	0	0	0	0	8	Calm			
08	30.8	0900	bc	5	0	0	6	6	WNW	1		
08	30.8	1200	bc	5	0	0	8	7	WNW	3		
08	30.8	1500	b	0	0	0	0	7	WNW	1		
08	31.8	0800	c	7	0	2	–	7	NNW	3		
08	31.8	1800	o	8	8	–	–	6	NNW	1		
09	1.9	0745	b	1	0	0	1	7	NNE	2		
10	1.9	0930	c	7	3	4	–	7	NNE	1		
01	1.9	1400	c	6	8	–	–	7	N	1		
01	1.9	1600	bc	5	0	0	9	7	N	1		

25 August is like 23–24 August allowing for new wind direction probably bringing slightly cooler upper air for greater growth of cumulus clouds. Night, however, brings

26 August, a big change of wind together with medium cloud type 2, which almost certainly means a front bringing in a new air mass. That is a local change of weather from the fair to the fine type again associated with high pressure. With your back to the north-east wind, you see, the pressure is higher upon your right hand, that is to the north-west of Britain, which usually means more settled weather.

27 August is a day of fine weather type just like 23 August except for the opposite wind direction and the temporary appearance of low clouds in the morning. At first, you see, when stirred by the heat of the sun on the ground, the air is still damp enough to form clouds as it rises under an 'inversion' layer. Later, however, it has become so warm and therefore relatively so dry that the condensation level cannot be reached. So the weather is cloudy at first but fine in the afternoon. As usual here with easterlies we notice slight haze from the continent too.

28 August is similar.

29 August shows the ideal weather of a high-pressure centre.

30–31 August. Observations suggest a feeble warm front, for with wind now from a north-westerly quarter the high-pressure centre must have shifted south to bring us into a sea air current again. 31 August is duly cloudy all day.

1 September is more like 20 August which showed polar-air weather, so perhaps yesterday's front was not really warm but occluded.

You see here how we need charts to show our air's history. Later in this book we shall have them; but for the present we learn all we can from our own local observations, however parochial. We must remember the proverbial wisdom of shepherds, farmers and fishermen.

5. FAIR WEATHER TYPE

Place no.	Date 1936	Time G.M.T.	Weather letters	Cloud				Vis.	Wind		Temp.	
				T.	L.	M.	H.	V.	D.	F.	F.	C.
01	2.9	0800	c	6	0	0	6	7	WSW	4	70·8	22
01	2.9	1200	c	7	0	0	7	7	SW	4	74·0	23
01	2.9	1330	c	6	4	8	8	7	SW	4	75·8	24
01	2.9	1630	bc	5	0	3	0	7	SW	3	73·7	23
01	2.9	2120	bc	4	0	0	9	6	SW	2	64·8	18

2 September is like 15 August but with more upper cloud, due perhaps to a weak or distant front. You notice the same

weather bc;

low cloud type 1 or 4 which means small cumulus flattening out, becoming 0 in the evening;

visibility 6–7;

wind direction south-westerly;

day temperatures in the seventies F., twenties C.

That is our early autumn fair weather type.

6. RAINY WEATHER TYPE

Place no.	Date 1936	Time G.M.T.	Weather letters	Cloud				Vis. V.	Wind		Temp.	
				T.	L.	M.	H.	V.	D.	F.	F.	C.
01	3.9	0800	o	8	5	–	–	7	SSW	2	66·2	19
18	3.9	1100	o rr	8	7	–	–	6	S'W	3		
18	3.9	1215	c	7	7	7	–	7	SSW	4		
01	3.9	1430	c	7	5	–	–	7	SW'S	4	69·3	21
01	3.9	1540	o $r_0 r_0$	8	5	–	–	7	SW'S	4		
01	3.9	1600	c $r_0 r_0$	7	7	2	–	7	SW'S	4	68·9	21
01	3.9	1745	o r q	8	7	–	–	7	SW'S	5	66·7	19
01	3.9	1810	c	7	7	1	–	7	SW'S	4	66·6	19
01	3.9	1820	bc	4	2	0	9	7	SW'S	4		
01	3.9	2130	bc	3	0	1	–	7	SW'S	4	62·2	17
01	4.9	0630	c	7	5	1	9	7	S	3	62·0	17
01	4.9	0710	c rr	7	0	2	–	7	S	3		
01	4.9	0800	o rr	8	7	–	–	7	S	5		
01	4.9	0945	o/rr	8	7	–	–	7	SW'S	6	62·3	17
01	4.9	1200	c	7	7	–	–	7	SW'W	6	65·8	19
01	4.9	1430	c pr q	7	7	–	–	7	WSW	4	65·7	19
01	4.9	1630	c	7	5	7	–	7	WSW	5	63·2	17
01	4.9	1800	bc	5	5	5	–	8	WSW	5		
01	4.9	2000	c	6	5	5	–	8	WSW	5	61·9	17
01	4.9	2230	bc	4	0	5	–	7	WSW	5		

3 September is a day of wet south-westerly weather. Conditions suggest a warm occlusion or front passing about midday. Afternoon weather also looks frontal though not so definite. Thus:

at 1100 the weather is overcast with continuous rain and ragged low clouds of bad weather (C_L type 7) while visibility is reduced to 4–10 km. (V = 6), improving

by 1215 when the rain has ceased and the clouds have begun to break (total amount now only seven-eighths of the sky), the medium cloud type (C_M 7) is still frontal but the visibility has improved (V = 7, which is good) and the wind has started to veer.

At 1430, however, it is still cloudy, and then

by 1540 it is raining again. The real clearance does not come until 1810, when the setting sun shines through the clouds as they change to the cumulus type and break to only four-eighths of the sky. It is one of those spectacular sunsets when the *under*-side of the upper clouds is lighted up and so appears to glow all orange against the darkening background.

4 September is similar but with a front appearing to pass about 0930 instead. Temperature is about its usual figure in strong south-westerlies (coming from sea with nearly this temperature). The actual range of temperature is small, owing partly to the cloudy sky.

At 0710 we see continuous rain for a couple of hours from medium cloud type 2 with ragged low clouds of type 7 while wind veers from southerly to south-westerly.

0945–1430 is cloudy.

From 1430 the weather is of the showery type, cloudy just at first but becoming fair. This suggests polar air, as the wind has veered from a southerly to a westerly quarter.

7. SHOWERY WEATHER TYPE

	Place no.	Date 1936	Time G.M.T.	Weather letters	Cloud				Vis. V.	Wind		Temp.	
					T.	L.	M.	H.		D.	F.	F.	C.
	01	5.9	0630	c/PR	7	5	5	–	7	WSW	4	60·9	16
[a]	01	5.9	1030	c	6	3	–	–	8	W'S	4	65·8	19
	11	5.9	1115	c/pr t	7	3	–	–	9	W	4		
	12	5.9	1130	c phr	7	3	–	–	9	W	4		
	14	5.9	1200	c PR h_0	7	9	–	–	9	W	4		
	15	5.9	1245	c pr tl	7	9	–	–	9	W	4		
	15	5.9	1400	c pr	7	3	–	–	9	W	4		
	13	5.9	1610	bc pr	5	3	6	8	9	W	3		
	01	5.9	1715	c PR	7	9	–	–	7	W	3		
	01	5.9	2000	o r_0	8	o	7	–	7	W	3	58·8	15
	01	6.9	0800	c	7	5	–	–	7	NW'W	4	57·0	14
	01	6.9	0850	bc	5	o	5	–	7	NW'W	4	60·7	16
	01	6.9	0915	bc	2	1	3	1	7	NW'W	4	62·0	17
	01	6.9	0940	bc	2	2	5	o	7	WNW	5	63·8	18
	01	6.9	1100	bc	5	4	7	4	9	W'N	5		
	01	6.9	1230	c	7	4	7	–	9	W'N	4	64·7	18
	01	6.9	1340	c	6	4	7	–	9	W	4	64·9	18
	01	6.9	1420	o	8	4	2	–	9	W'S	3	64·2	18
[b]	01	6.9	1540	o r_0	8	o	2	–	8	WSW	4	62·9	17
	01	6.9	1650	o	8	o	2	–	8	SW'W	3	61·9	17

[a] Cumulonimbus impressively large for so early in the day.
[b] Medium clouds coming from W.

5 September shows plain deep showery polar air, evidently after an early morning cold front passage. Normal features are the cumulonimbus clouds giving showers with thunder and hail in the afternoon. Apart from the rain the visibility is excellent. In heavy rain, of course, it is poor, and should, strictly speaking, be reported so.

6 September shows frontal characteristics of cumulus spreading out by midday while medium cloud spreads above it; warmer (wetter) air aloft to herald a warm front.

IV

AIR THERMODYNAMICS

1. WET AND POTENTIAL TEMPERATURES

You will have noticed that not only fair-weather clouds but even thunder-storms may be formed as the air goes along. Now, to repeat first principles from Chapter I, 'our analysis of how properties change as the air goes along is largely a matter of thermodynamics, the theory of heat flow. That is part of physics. It speaks of temperature, moisture, pressure—the ABCD of our language. If theory and practice are to talk the same language, then we must express observations in the same terms. Moisture, pressure and temperature can be measured. All that remain are precipitation, clouds and visibility to be described in figures too.'

That is what we have now done. The time has come for us to see just how moisture, pressure and temperature are related to one another and so to the weather.

'The weather', to quote our first chapter again, 'is simply a matter of water changing its state. But that means gain or loss of latent heat, for which Nature has a rationing system.' That is our first law of the atmosphere. Air with its full ration of water vapour is SATURATED, so the full ration itself may be called the saturation moisture content (s.m.c.), of which the *actual* moisture content (m.c.) is a percentage called the relative humidity (r.h.).

How can they be altered?

One way is for temperature but not pressure to change. That is familiar enough on the ground any day. What then happens to the r.h.? Over dry ground the *actual* m.c. may not change, but the SATURATION m.c. depends on the temperature, which does change. So their percentage *ratio* changes. That is the r.h. How is the weather affected? Chiefly in VISIBILITY. We shall discuss that later on.

How else can r.h. be altered?

Suppose the air rises. Higher up it has less air above it to press down upon it, and so it expands. Expansion uses up energy. Without any outside source of energy the air has to draw upon its own private heat, and so it cools down. This way of cooling is given a name. We call it ADIABATIC.

Descending air likewise is warmed. A simple law then relates the temperature with pressure and so with height. The rate of change of temperature with height is the so-called ADIABATIC LAPSE RATE. With this, you see, the rising air's temperature is given by just two things together, either the *height* with ground temperature (or the INITIAL TEMPERATURE, say just over the ground), or else the PRESSURE and POTENTIAL TEMPERATURE (which means the temperature reduced to, or taken up at, some standard pressure). Of those and the s.m.c., any two will determine the third; so at the cloud base or saturation level the actual m.c., being saturation m.c., will determine the height in terms of the air's potential temperature or ground temperature. That gives the height of the base of the clouds.

The air on the ground, whenever sufficiently wet or warm, and therefore buoyant, must rise. Like all things on earth it has what we call gravitational potential energy by virtue of its height above ground. How much? Just the standard gravity (g) times how high up it is.

In the standard JOULE units of physics this amounts to almost exactly 1% per metre. As it has the effect of cooling the air 1% of a degree Centigrade as it rises, it determines the dry adiabatic lapse rate as 1% of a degree Centigrade per metre, or 10° per km.

Rising air, for instance, initially 10° C. warmer than saturation temperature (dew-point), can rise a whole kilometre before forming clouds. In fact as the dew-point itself will have been slowly falling, the cloud base must be even higher—about 4000 ft. That is just 400 FT. MULTIPLIED BY THE INITIAL EXCESS OF TEMPERATURE OVER DEW-POINT in degrees Centigrade. That is a useful rule. If temperature over the countryside is fairly uniform, while m.c. is even more nearly uniform, then so must the cloud height be. You will often have noticed it so. Over the sea it is more obvious still. The bases of cumulus clouds for some distance inland all appear at about the same height above sea-level, being reached then by a shorter ascent from the hilltops than from the valleys because the cool hill air has less excess of temperature over dew-point.

In clouds the air receives latent heat from condensing moisture. Above the level of their base it therefore cools off more slowly with height. This reduced rate is the so-called saturated or WET ADIABATIC lapse rate. With this the rising air's temperature is determined not by ground temperature with height above ground but by cloud-base temperature together with height above cloud base.

Have you ever been near the equator? Even if you have never yet 'crossed the line' nor been in the tropics, at least you will know how notoriously humid the equatorial regions are. Yet not even residents always realise what that humidity is. 'Tell me,' one of them says to the meteorologist, 'Does it actually ever reach one hundred per cent?'

'Tell me,' one of them may have said to the doctor with great agitation, 'Have I got blood pressure?' This question has something in common with the other, equally futile one: it does not quite represent what is wanted. Blood pressure may be important; yet merely to have blood pressure is nothing: it is only *abnormal* blood pressure that counts. So is moisture important; yet mere saturation (100% r.h.) is nothing but fog, such as you may find anywhere in the world or up in the sky as clouds. No such percentage humidity figures can distinguish the special humidity of the tropics. After all, what is this percentage? It is only relative to how much moisture the air *can* hold, which you still may not know. We should think not of relative but of absolute things. It is for *absolute* humidity that the tropics are so notorious.

How is it to be described? Surely the easiest way is to say WHAT WEIGHT OF MOISTURE IS HELD IN UNIT WEIGHT OF AIR. That is the proportion (percentage, say) by weight, not just of how much moisture the air *can* hold (which you still do not know) but *of the whole air itself* which you do know. Meteorologists call this the air's HUMIDITY MIXING RATIO. In this book, however, we shall take MOISTURE CONTENT or ABSOLUTE HUMIDITY to mean the same thing.

You know how the air is a mixture of gases—nitrogen nearly 80%, oxygen nearly 20%, and so on. One of them is water vapour. But of this the air cannot hold any more than a certain percentage by weight, depending on pressure and temperature. In the tropics, for instance, all over the warmest seas at about 80° F. or 27° C., the air can take up about 2% of its own weight in water vapour. Its absolute humidity then, you see, is 2% (of the air as a whole) while at saturation its *relative* humidity is 100% (of this absolute amount). $2\frac{1}{2}$%, however, is rather rare, and with 3 you could hardly live. In the British Isles the highest is about 1%, such as in autumn south-westerlies or on a 'close' summer day when fog at dawn is enough to show you that this 1% moisture will saturate the air at about 15° C. or nearly 60° F. At freezing-point (0° C.), on the other hand, the air cannot hold more than about a *third* of 1% moisture. If then warmed up during the day to nearly 60° F., at which 1% is what it *can* hold, *relative* humidity falls to about 33%. Water can then evaporate freely into the air. Notice, however, that what determines *how fast it evaporates* is not purely this 33% *ratio* of actual to possible value, but rather their *difference*, which is the difference between $\frac{1}{3}\%$ and 1% of the air. Such air would feel very dry and mild in this country, particularly in spring.

On a close summer day, on the other hand, starting saturated with 1% vapour at nearly 60° F., the air may be warmed up to 88° F., at which it can hold 3%. Again then with the same *ratio* of *actual* to *possible* moisture the r.h. is 33%; but water dries up much faster now because the *difference* is no longer only $\frac{2}{3}$ of 1% but is now the whole difference between 1% and 3%, that is 2% of the air. So the rate of drying of water illustrates how we learn more from absolute than from mere relative humidity. Consider your comfort, too. Your body temperature, nearly 99° F., is 37° C. Easier to remember, perhaps, is how far it is above the *absolute* zero $(-273°$ C.$)$. $273° + 37° = 310°$. The air in hot countries may have this same temperature, so that without some special cooling device your body could not possibly lose any heat to the air. As heat is nevertheless being generated inside you, some special device is essential. Nature's own simple device is perspiration. How does it work?

When water evaporates into the air it takes up heat, but in LATENT form that is useless for warming the air. As the liquid, or *anything wet* that is freely exposed to the air, thus loses heat and cools down, the air feeds heat back to it out of its ordinary (not latent) store that only depends on the temperature difference. The cooling, of course, goes on until these opposite heat flows balance each other, when it stops at what we may call the WET TEMPERATURE. Meteorologists call it the *wet-bulb* temperature; but let us not bother with bulbs just yet.

Much of the British Commonwealth lies between desert and deep sea so that it only has two kinds of air. One kind comes from the desert extremely dry, say with only $\frac{1}{3}$ of 1% water vapour, while the other kind is from the sea which gives it 2% vapour. Perspiring freely exposed to the desert air at 37° C., then, your skin is cooled by evaporation according to the big difference between $\frac{1}{3}\%$ *actual* and 4% moisture content *possible* at first. But at 31° C., which is 88° F., only 3% is now possible, so that the difference is less and so the evaporation is slower. Heat meanwhile is being fed back according to the temperature difference of 6° C. from the air.

At 15° C., at which the air cannot hold more than 1% vapour so that the *actual* and *possible* moistures now only differ by $\frac{2}{3}$ of 1%, evaporation is so slow as to balance the now large (37–15° C.) temperature difference effect. So you have reached your wet temperature, which is no higher than that of the English summer dawn. Though parched, you feel cool. Actually you are sweating briskly but never notice it because equally briskly is it evaporating. You only detect it indirectly by thirst.

Perspiring equally freely, exposed, on the other hand, to deep-sea air at 37° C., your skin is cooled by evaporation according only to the small difference between *actual* 2 and *possible* 4% moisture at first. The cooling stops at wet temperature 28° C. That is little below 37°, but then there is only the difference between *actual* 2 and *possible* 2½% vapour into which to evaporate. Whilst perspiring perhaps no more than before, you stream with sweat because it cannot dry away fast enough.

Humidity, then, is measured with a pair of thermometers. One of them has a dry bulb, the other a wet bulb (wrapped in wet muslin). When steady, the wet bulb is taking heat from the air just as fast as it gives latent heat to the evaporating water.

To what is the flow of heat from the air proportional? Simply to the DEPRESSION OF WET-BULB TEMPERATURE BELOW DRY.

To what, on the other hand, is the flow of latent heat proportional? Simply to the flow of vapour itself.

To what, in turn, is the vapour flow proportional? Obviously to the vapour pressure difference between the wet-bulb air and the dry. That represents the depression of the outside air's ACTUAL MOISTURE CONTENT below WET-BULB SATURATION MOISTURE CONTENT.

But s.m.c., as you know, is already determined by the pressure and temperature themselves. Thus the wet-bulb s.m.c. is already determined by the pressure and wet temperature. That leaves only the air's actual m.c. unknown. So we can simply deduce it from the wet and dry temperatures. Dry 37° with wet 15° C., for instance, corresponds as we have seen to m.c. of the order of one-third of 1% of the air, while dry 37° with wet 28° C. determine m.c. about 2%.

Now suppose the air rises. We have seen that its DEPRESSION OF WET TEMPERATURE BELOW DRY is proportional to its DEPRESSION OF ACTUAL M.C. BELOW WET-BULB S.M.C. How do they all vary with height? In other words what are their lapse rates?

To determine the *lapse rate* of wet temperature depression we have only to know those of dry temperature, actual m.c. and wet-bulb s.m.c. That of dry temperature of the rising air, of course, is simply the *dry adiabatic* lapse rate. That of the actual m.c. is ideally *zero* as long as the rising air gains or loses no moisture. What about the wet-bulb s.m.c.'s variation with height? That simply means the rate of reduction of how much water-vapour the rising air can hold at wet temperature. That is the *rate of condensation* of surplus water from rising air saturated at wet temperature, which determines of course the RATE OF SUPPLY OF LATENT HEAT OF CONDENSATION, which we may call R. Now in clouds, as we saw in an earlier paragraph, the air receives latent heat from condensing moisture so that it cools not at dry adiabatic but at the (smaller) wet adiabatic rate. So R REPRESENTS THE DIFFERENCE BETWEEN THE WET AND DRY ADIABATIC LAPSE RATES. That, then, gives the wet-bulb s.m.c.'s variation with height. Subtracting the actual m.c.'s variation with height, which is ideally zero, you see that the difference between wet and dry adiabatics gives the lapse rate of the depression of actual m.c. below wet-bulb s.m.c. But that is the lapse rate of depression of wet temperature below dry. So the WET-TEMPERATURE LAPSE RATE IS THE WET ADIABATIC.

Read that argument again to make sure. Although it is not the proper thermodynamical argument, it conveys the idea, without mathematics. For air which is saturated, of course, the conclusion is more obvious, as the wet and dry temperatures are the same anyway, and you already know that the dry temperature then has wet adiabatic lapse rate.

The warmer and moister the air, the more latent heat is set free as it rises, so the bigger is the difference between wet and dry adiabatic lapse rates. In tropical air, for instance, while dry rate is 10° C. per km. of height, which is just over 15° C. per mile or per 5000 ft., the wet rate is more nearly 10° C. per mile.

Evaporation can cool the air to wet temperature. Having saturated the air at wet temperature, it must stop. Being thus the minimum temperature to which the air can be cooled by evaporation of water, the wet temperature is unaffected by this evaporation. NO EVAPORATION NOR CONDENSATION OF WATER INTO THE AIR CAN AFFECT THE WET TEMPERATURE.

Meanwhile we have just seen that wet temperature has its own standard (wet adiabatic) lapse rate, by which we may therefore reduce it to standard pressure just as we do with dry temperature. What shall we call the result? Just as the dry temperature, when reduced to standard pressure, is called potential temperature, so the wet temperature, when reduced to standard pressure, may be called WET POTENTIAL TEMPERATURE (w.p.t.). Neither evaporation nor condensation of water nor any ascent nor descent can therefore alter w.p.t., the air's most conservative property. Next best are dry potential temperature (d.p.t.) and m.c., being unaltered by any adiabatic ascent or descent as long as the air is *not saturated*. Ascent and descent, however, may *redistribute* them. MIXING, for instance, evens them out. STIRRING, of course, is a way of mixing. That is TURBULENCE, of which you will hear much.

Just as the dry temperature of rising air is determined alone by pressure and d.p.t. with standard dry adiabatic lapse rate, so its wet temperature is determined by pressure and w.p.t. with wet adiabatic. Other things being equal, for instance, all air rising from the earth's surface will keep the w.p.t. that it has picked up on the surface, with which its wet temperature at any height can duly be deduced from the pressure at that height. The temperature at the top of a cumulus cloud then simply depends on how high it is.

Convection, however, is a turbulent process which redistributes w.p.t. No longer are other things equal. Wet rising air is mixed with drier surroundings. What happens? Suppose the old tropical sea air with 2% moisture is thoroughly mixed with dry desert air whose moisture content is only $\frac{1}{2}$%. What stirs them? It is really the buoyancy of the rising sea air that is just being heated by the ground. Its buoyancy, of course, is the resultant of its own weight with the weight of the air it displaces. Imagine them on a pair of scales. The buoyancy then is the up-thrust upon the lighter scale pan as they fail to balance.

Next suppose that the beam which holds the pans is pivoted not at the middle point, but at wherever it will in fact balance. If the weights are respectively 1 lb. and 2, then this point must be twice as far from the light end as from the heavy. In short it is not the middle or *mean* point but the *weighted mean* point. On a 3 ft. beam, for example, it must be 1 ft. from the 2 lb. end whilst 2 ft. from the 1 lb. end. Its distance along the beam then is simply reckoned by dividing the *total weight* (3 lb.) into the *total weighted distance* (1 lb. multiplied by the 1 lb. weight's distance, plus 2 lb. multiplied by its own distance). Distances to the right or left are distinguished as plus or minus.

Now when about 1 km. depth of sea air, which has 10% of the whole atmosphere's pressure or weight, and of which just 2% is moisture, is mixed with about 2 km. depth of dry desert air having twice as much weight, of which in turn only $\frac{1}{2}$% is moisture, you deduce the total amount of moisture simply as the TOTAL WEIGHTED M.C., which is $\frac{1}{10}$ of 2% $+ \frac{2}{10}$ of $\frac{1}{2}$%, that is, altogether, $\frac{3}{10}$ of 1%. Meanwhile what is the total weight or pressure of these two masses of air? It is three tenths of the whole atmosphere's. Dividing the total weight then into the total weighted m.c. you find the WEIGHTED MEAN m.c. = 1% of the air. That is the m.c. of our mixture. At our body temperature of 37° C. this mixture could hold 4% while its actual m.c. is 1%. So its r.h., their ratio, is 25%. Wet temperature is then about 22° C. or 70° F., at which you might feel fairly comfortable, not too hot, nor too parched. Cumulus clouds in the sea air with w.p.t. 28° C.

would be dried down to the new w.p.t. 22° C. To keep their original temperatures then the tops would not be so high. *That* is what happens to them. The drier the upper air is, the more marked may this effect be.

The longer the convection goes on, the more air is fed from the ground, which tends to bring the upper-air w.p.t. to the ground- or sea-surface value. That is in fact partly how an air mass is made. First recognised by *horizontal* uniformity, it tends to acquire also vertical uniformity of w.p.t. W.p.t. is therefore the property by which it may best be defined. It is one of the simplest functions of our weather elements that almost uniquely defines an air mass.

Up to now we have spoken of lapse rate as rate of variation with height of rising air—by virtue of its rising. More generally, however, it means variation with height not following the air up but just measured over any fixed place, regardless of whether the air is rising or not. CONVECTION is then THE RISING OF AIR THROUGH ENVIRONMENT WHOSE TEMPERATURE LAPSE RATE EXCEEDS ADIABATIC.

2. CONVECTION

You have already seen how the air's m.c. (moisture content) and d.p.t. (potential temperature) determine the height of cumulus cloud base. What was the rule? Approximately the height is 400 ft. multiplied by the initial excess of temperature over dew-point in ° C. That is admittedly not quite in terms of d.p.t., nor entirely in metric units, but you will remember it easily.

Above cloud base the air is no longer dry but saturated, with wet adiabatic lapse rate, so the temperature is determined not by ground temperature with height above ground, nor by d.p.t. with height above standard-pressure level, but by cloud-base conditions together with height above cloud base, or more simply by HEIGHT AND W.P.T.

'Other things being equal,' we said, 'air rising from the earth's surface will keep the w.p.t. that it has picked up on the surface, with which its wet temperature at any height can duly be deduced from the pressure at that height. The temperature at the top of a cumulus cloud then simply depends on how high it is.' Naturally the air stops rising when at the same upper-air temperature (u.a.t.) as its surroundings. Of the three factors, (1) w.p.t., (2) height of cloud tops, and (3) u.a.t. of environment, therefore, any two will determine the third.

The drier the air, for instance, the less latent heat of water it has, the lower is its w.p.t. and so the lower are its cloud tops, as we have already

	Place no.	Date 1936	Time G.M.T.	Weather letters	Cloud T.	L.	M.	H.	Vis. V.	Wind D.	F.	Temp. F.	C.
	01	7.9	0700	o/r	8	7	–	–	7	W'S	4	62·2	17
	01	7.9	0900	b	1	1	0	0	7	WNW	7	64·0	18
	01	7.9	1100	bc	5	2	0	0	8	W	6		
	01	7.9	1530	bc	5	3	6	0	8	W	6	64·5	18
[a]	01	7.9	1540	c	7	8	–	–	8	W	6	63·5	18
	01	7.9	1550	c	7	9	–	–	8	W	6		
	01	7.9	1615	b	2	4	6	0	8	W	6		
	01	7.9	1800	bc	5	4	6	0	7	W	6	60·5	16
	01	7.9	2050	bc	5	4	6	0	8	W	6	58·7	15
	01	7.9	2120	c	7	–	–	–	8	W	7	58·5	15
	01	8.9	0700	o rr	8	7	–	–	6	W	5		
	01	8.9	0745	c rr	7	7	–	–	6	W	5		
	01	8.9	0800	c	7	9	–	–	6	W'N	5	56·2	13
[b]	01	8.9	0810	c pr	7	9	–	–	6	W'N	5	57·0	14
	01	8.9	1040	o	8	8	–	–	7	W'N	5	59·5	15
	01	8.9	1150	c	7	9	7	3	7	NW'W	5	61·5	16
	01	8.9	1330	c	6	2	6	3	7	NW'W	6	64·3	18
	01	8.9	1500	c	7	2	3	–	7	NW'W	5	62·9	17
	01	8.9	1700	c	7	4	3	–	7	WNW	4	62·0	17
	01	8.9	2020	b	1	4	0	0	7	WNW	3	59·6	15
	01	9.9	0700	o m	8	5	–	–	4	Calm			
	01	9.9	0915	o m	8	7	2	–	4	ENE	1	58·5	15
[c]	17	9.9	1030	o m	8	7	1	–	4	ENE	1		
	16	9.9	1200	c m₀	7	0	7	–	5	E'S	1		
	01	9.9	1350	c	7	0	7	–	6	SE	2	65·7	19
	01	9.9	1415	c pr	7	2	7	–	6	SE	2	65·5	19
	01	9.9	1615	c	6	1	7	9	7	SE	3	65·7	19
	01	9.9	1800	bc	3	3	7	–	6	SE	2	63·6	18
	01	9.9	1850	bc	3	3	0	0	6	SE	2	62·8	17
	01	9.9	2130	bc/b	4	0	6	0	6	SE	2	58·2	15

[a] Cloud tops moving from WNW. Relative motion of cloud bases is from WSW.
[b] Four showers between 0810 and 1010.
[c] Bar. rising.

seen when allowing for mixing. The lower the u.a.t., on the other hand, the colder the upper air is, the warmer is the rising air relative to it, and so the higher are the cloud tops.

7 September illustrates these. After early morning rain as on 5 September we are in clear polar air. Almost cloudless at 9 a.m. the air rising over sunny ground at about 18° C., being probably 5° C. above dew-point, makes cloud-base height about 5 times 400 which is 2000 ft. In 2000 ft. at the dry rate of 10° C. per km. it will have cooled to 12° C. Meanwhile at wet rate the wet temperature will have fallen from w.p.t. (about 15° C.) to 12° C. also. With wet and dry temperatures now the same, the air must rise saturated in the form of cumulus clouds until equal in u.a.t. to the polar upper air already there.

Sure enough, cumulus clouds (C_L2) are reported at 11 a.m. C_L1 was actually developing into the larger type 2 all the time until 1330, and then for an hour their respective amounts remained constant. If you look up their codes in Chapter II you will see that the tops of type 1 are generally at 1–2 km. while those of type 2 are at 2–4 km. Apart from the occasional cumulonimbus (C_L types 3 and 9) reported this afternoon, perhaps over-estimated in size as no shower was noticed, most cloud tops must have been at 2–3 km.

Then we can infer the u.a.t. from the w.p.t. 15° C.; for in 2–3 km. at wet rate of 10° per mile or about 6° per km. the wet temperature must have fallen to about freezing-point, 0° C. So 2–3 km. must have been the freezing level in the polar upper air already there. But on 5 September the clouds at that height were still growing briskly with almost certainly no greater w.p.t. than to-day, so the upper air must have been colder. How can two polar air masses differ in u.a.t.?

Polar air, you must remember, is born over ice, which makes its u.a.t. very low. When it comes off the ice on to open water or land above freezing-point which thus warms the air at the bottom, it naturally starts overturning with clouds feeding into it the w.p.t. from the surface. Over the sea, for instance, where wet temperature follows sea temperature, the whole air mass tends to acquire w.p.t. nearly equal to the temperature of the sea underneath it. If it rapidly crosses the sea isotherms from cold to warm, then it lags behind the sea temperature, so that a straight burst of arctic air over Britain in winter with sea temperature 5–10° C. may still have w.p.t. below zero (° C.). That is our coldest polar sea air in this country. But if it has had time to curve round so as to run with the isotherms, then it may almost catch up on sea temperature, so that ex-polar air coming round into Britain from the south-west in summer may have w.p.t. up to 15° C.

Wind to-day was strong, and veered with height in the normal way as you notice from the remarks. Total cloud amount grew to seven-eighths of the sky *as seen from the ground*, though if seen from a great height or on a map the cloud-filled up-currents and compensating down-currents of air would probably be about equal in horizontal extent.

Every cumulus cloud, of course, marks where the air is going up. The air flows into its base and out of its top—or rather, out *with* its top, and downwards all round it. This descent of air all around the cloud will tend to suppress weaker neighbours (smaller cumulus clouds), so that convection is an example of survival of the fittest.

8 September is similar, after rain at first. **9 September** is rather similar but has the opposite wind as the ex-polar air mass settles down with high pressure over Europe. Coming from the smoky side of London it brings thick haze, while with overcast sky the sun cannot warm and so dry out the air till midday. So we introduce the subject of visibility.

3. VISIBILITY

	Place no.	Date 1936	Time G.M.T.	Weather letters	Cloud T.	L.	M.	H.	Vis. V.	Wind D.	F.	Temp. F.	C.
	01	10.9	0640	o m rr	8	7	–	–	4	SE	2	59·0	15
[a]	01	10.9	0745	o r$_0$r$_0$	8	7	–	–	4	SE	2		
[b]	01	10.9	1000	o m$_0$	8	7	–	–	5	SE	·2		
	01	10.9	1100	o	8	3	7	–	6	SE	2		
	01	10.9	1130	c	7	3	7	9	6	SE	3		
[c]	01	10.9	1140	o	8	7	2	–	7	SSE	3		
	01	10.9	1320	o	8	7	–	–	7	S'W	3	66·6	19
[d]	01	10.9	1330	c u v	7	9	–	–	7	S'W	2		
	01	10.9	1630	bc	2	3	5	3	7	S'W	3	66·8	19
	01	10.9	1830	b	1	5	5	0	7	S'W	3	63·5	17
	01	11.9	0515	b	0	0	0	0	6	E'S	2	58·5	15
[e]	01	11.9	0530	b w	1	0	3	1	6	E'S	3		
	18	11.9	1200	b	1	1	0	1	8	—	3		
	01	11.9	1800	b	0	0	0	0	6	S	3		
	01	12.9	0620	c/bc	7	7	5	6	7	SE'S	3	60·8	16
	01	12.9	0730	o	8	0	7	7	7	SE'S	3	61·4	16
	01	12.9	1130	o r$_0$r$_0$ > rr	8	0	2	–	6	SE'S	3	65·5	19
[f]	01	12.9	1450	c	7	2	7	–	6	SE'S	3		
	01	12.9	1530	c	7	0	7	9	6	SE'S	3		
	01	12.9	1630	c	7	0	0	9	6	SE'S	3		
[g]	01	12.9	1800	c u PR	8	0	8	–	6	—	3		
	01	12.9	2000	c pr	7	0	9	–	6	SSW	2	65·1	18
[h]	01	13.9	0730	o m$_0$	8	5	–	–	5	WSW	2	63·0	17
	01	13.9	0920	o u$_0$	8	5	2	–	5	WSW	2		
	01	13.9	1115	c m$_0$	7	3	0	2	5	WSW	2		
	01	13.9	1250	bc	5	2	6	8	6	WSW	3	69·5	21
	01	13.9	1330	c	6	3	0	3	6	W	2	70·2	21
[i]	01	13.9	1720	bc	4	3	5	8	6	NW'W	2	68·3	20
	01	13.9	1915	b	0	0	0	0	6	NW'W	1	64·1	18

[a] Rain decreasing, after being heavy.
[b] Cloud height only 600–700 ft. Upper clouds moving from SW.
[c] Aerial perspective (see text) almost disappears at noon, reappearing at 1230.
[d] A.P. again disappears for half an hour.
[e] Medium clouds moving from SE.
[f] Rain ceases at 1330. Medium clouds moving from SSW.
[g] Rain for 20 minutes. Medium clouds still moving from SSW.
[h] Height of clouds 400 ft., rising to 700 by 0920.
[i] Medium clouds moving from SW'S.

10 September shows signs of a front at midday. There is the normal kind of contrast between conditions ahead of it and behind it, but some subtlety about the time of its arrival. The new wind direction, west of south, appears first at upper cloud levels and then at the ground some time between 1140 and 1320. But it is of visibility that we have more to say.

AERIAL PERSPECTIVE is the familiar sight of successive lines of hills or woods in the distance looking successively lighter grey as the air particles which scatter the light are seen in greater and greater depth. This effect vanished at noon and again at 1330 together with what is called 'unusual visibility', which is perhaps more easily recognised than described. Not only the uneven lighting but also the arrival of the new air mass, presumably having a different concentration of particles scattering the light, comes under suspicion as the cause. It suggests noon as the time of arrival of the main front at ground-level, possibly followed by a secondary one at 1330. The cloud clearance lags behind them, as with a cold front.

The basic principle of daytime visibility is that it is represented by THE DISTANCE AT WHICH OBJECTS BECOME INDISTINGUISHABLE FROM THEIR BACKGROUND. What we may call its standard value, say V, is for black objects against the sky. The principle then is that there is a certain minimum detectable relative difference between the intensities of light scattered by air particles in unlimited depth (the sky) and by those in depth V (in front of the objects). By simple theory of light scattering, this difference is $e^{-V\sigma}$ where e, the mathematical constant, is 2·718..., while σ, the so-called extinction coefficient, is determined by the colour of the light and by the CONCENTRATION and SURFACE AREAS of molecules, liquid droplets and solid particles in the air.

Taking $e^{-V\sigma} = 2\%$, for instance, we find $V = \dfrac{3\cdot9}{\sigma} \simeq \dfrac{4}{\sigma}$. The air's so-called

OPACITY, then, being $\dfrac{1}{V}$, is approximately $\dfrac{1}{4}\sigma$. Pure air molecules' contribution to this opacity is found to limit horizontal visibility over the ground to something of the order of 250 km. or 150 miles. At that distance a mountain would appear so pale blue-grey as to be well-nigh indistinguishable from the sky behind it. A dazzling snow-cap might show it up, but then we should no longer have the standard condition of a black object against the sky. The effects of colour, in fact, form a long story that need not be told here.

Smoke or dust particles contribute their own opacity in proportion to their average concentration and surface area, while hygroscopic particles also absorb and duly dissolve in some of the air's moisture to form so-called NUCLEI which grow into droplets whose concentration and surface areas are largely determined by the air's HUMIDITY. We shall say more about this subject under the heading of FOG.

11 September is a fine day like 24 August with merely a wind from the continent. Cool and almost cloudless, with dew and slight mist (visibility figure 6) at first, the air becomes clear with only a trace of cumulus, showing that it has become relatively dry and/or has an 'inversion' formed as a result of high pressure over Europe. While the wind at the ground is only just south of east, the medium clouds come from the south-east. That shows a veer of wind with height, probably only the normal one up to less than 1 km., beyond which the wind direction would hardly alter. With your back to the south-east wind you would have higher pressure on your eastern hand, that is over Europe. So simple an observation as this of upper cloud's motion is often most helpful in showing you at a glance where the weather is really coming from. Though not an infallible rule, it is often much better than merely watching the lowest clouds.

12 September is in many ways like 10 September. The change of wind and the heavy rain falling from upper clouds alone suggest the passage of a front in the evening. Again you notice the new wind appearing first at the upper cloud levels and then at the ground as the weather comes in from its new direction, no longer from Europe but from the Atlantic. The front, in fact, is just the dividing line between continental and maritime air. While the weather type characteristic of ex-tropical sea air is cloudy, or of ex-polar sea air is showery or fair, that of dry land air is fine. Yesterday's was the fine type; this morning's was the frontal type of continuous rain (with characteristic sequence of high cloud types 6, 7, and medium cloud types 5, 7, 2, which are followed by low clouds only upon arrival of the new air at low level), and then this evening is showery as in polar air.

13 September is rather like 9 September except for opposite wind and less upper cloud. It is also like 18 August. The wind change, a veer at ground-level before any change observed in the upper cloud motion, suggests a cold front.

4. FOG

	Place no.	Date 1936	Time G.M.T.	Weather letters	Cloud				Vis. V.	Wind		Temp.	
					T.	L.	M.	H.		D.	F.	F.	C.
	01	14.9	0615	b	2	0	3	5	6	WSW	2	56·9	14
[a]	01	14.9	0720	o f	8	5	–	–	3	WSW	2	58·7	15
	01	14.9	0830	o m	8	5	–	–	4	WSW	3	60·1	16
	01	14.9	0930	c	6	1	0	2	6	WSW	3	63·8	18
	01	14.9	1000	bc	4	2	0	2	6	WSW	3		
	18	14.9	1145	bc	4	3	0	2	7	WSW	3		
	01	14.9	1215	bc	3	2	0	2	8	WSW	3	67·0	19
	01	14.9	1450	bc	5	3	6	–	8	SW'W	3	67·7	20
	01	14.9	1500	bc pr$_0$	5	3	6	–	8	SW'W	3		
	01	14.9	1605	c PR	7	3	6	–	7	WSW	3		
[b]	01	14.9	1650	bc/pr	5	3	6	8	7	WSW	3	63·3	17
	01	14.9	1740	bc	3	3	6	0	7	WSW	3	62·0	17
	01	14.9	1810	bc	4	4	6	0	7	WSW	3	61·2	16
	01	14.9	1830	bc	5	3	6	8	7	WSW	3	61·4	16
	01	14.9	2000	bc	5	–	6	–	7	WSW	3	60·1	16

[a] Fog thickness variable.
[b] Rainbow seen.

14 September is rather like 7 September but with lighter wind and with fog or low stratus cloud covering the observation station for an hour or two after sunrise. This is now happening with temperatures only a few degrees below 15° C., which is nearly the temperature of the sea from which the air comes. The sea at this time of the year will only just have passed its 'midsummer' with its annual maximum temperature, whilst inland the air is cooling down for the autumn, reaching night minima of 13° C. quite easily. How is visibility affected?

The molecules of the curved surface of a drop of water, being more exposed, so to speak, than those of a flat surface, must escape or evaporate more, thus requiring more than the air's normal s.m.c. to balance the loss. This balance is unstable, because drop growth promotes itself by reducing the curvature and so reducing the rate of evaporation. *Dissolving* anything in the drop will slow down the evaporation, the more so the further the drop shrinks and the solution thus strengthens, until the actual air's m.c. (even below normal s.m.c.) suffices for balance. This balance is stable, because drop growth weakens the solution, thereby hastening the retarded evaporation and so causing shrinkage again.

Up to r.h. 100% this solution effect exceeds the curvature effect, so that as long as there are any hygroscopic particles floating in the air to form a solution at all the air will support some droplets, whose size will depend upon the r.h. whilst determining at the same time the opacity or *visibility*, which thus deteriorates as r.h. rises.

How far can r.h. rise? At 100% the actual m.c. has reached s.m.c. Over a flat water surface this is the highest m.c. possible in the form of *vapour*. Any excess must be liquid water, condensed in drops as fog. Its distribution among them naturally gives them an average size increasing with the excess of actual m.c. over s.m.c., or rather of actual water content (vapour plus liquid) over saturation *vapour* content. Thus as the actual m.c. is increased, or as the temperature and s.m.c. are decreased, visibility is reduced. That is also what happened with r.h. below 100%, but now it is due not to droplets but to bigger water drops having a different rate of change. S.m.c. changes quickly with temperature if high, but only slowly if low, so that by cooling the air the same number of degrees we condense more moisture and thus produce a denser fog in mild weather than in cold. Moreover, after changing steadily with r.h. rising up to 100%, the deterioration from mist to fog is relatively sudden. Owing to the curvature effect it cannot begin until m.c. slightly higher than normal s.m.c. is reached. That is not until r.h. has slightly exceeded 100%. But then because the balance is unstable the growth of the drops is sudden, producing a fog all at once.

Definite though these principles are, no fog is really easy to forecast. The simple idea that it starts when temperature has fallen to dew-point is complicated inland by the dew-point itself slowly falling as the air at standard thermometer level is dried out by the deposit of dew out of the air on the ground underneath, which is cooler. Moisture is also carried upward or downward by diffusion or stirring. The final dew-point, at which the fog really begins, is by no means always the same as the dew-point measured beforehand. In fact the estimation of this FOG-POINT involves considerable experience, often being near the initial dew-point not at thermometer level (just a few feet above ground) but averaged all the way up to the *final* level of the day's cumulus cloud base. That is the result of stirring into drier air higher up.

Still harder to estimate is the rate of cooling itself. Here are some of the factors:

THE GROUND MATERIAL AND ITS MOISTURE, a good conductor of heat being quickly replenished with heat from below.

THE AIR STABILITY, SUBJECT TO STIRRING BY WIND, the effect of turbulence being to even out the fall of potential temperature instead of leaving it all concentrated at the bottom.

THE AIR'S WATER VAPOUR, which sends back some of the heat.

CLOUDS, which send back most of the heat, according to their amount; and

CLOUD HEIGHT, since clouds which are low are warmer, and so send back more heat than high clouds.

Broadly, however, the principles can be made clear. Air that is cooled by the *sea* below dew-point, for instance, has SEA FOG, land complications being absent. Sea temperature is slow to change, so a sea fog is normally simply due to advection or bodily transport of air across the sea isotherms below dew-point. Advection fog, too, may occur over land, e.g. in mild air over snow that is being thawed by it, though even this has been debated. A variant is low stratus cloud, favoured by air-flow over land owing to the rough stirring. Air which is cooled over land by night-time radiation has RADIATION FOG when fog-point is reached. Clear skies, calm, dry air and poorly conducting ground surface favour the cooling; warm sea origin favours high fog-point. The season and air-mass type alone, in fact, are a good guide to the fog-point. To-day, for instance, we see how the maritime south-westerly air stream particularly favours this fog in autumn.

Other recognised types of fog are:

HILL (UP-SLOPE) FOG, such as a low-level place would see as cloud.

FRONTAL FOG, due (roughly speaking) to the mixing of warm wet air with colder dry air like the formation of clouds of 'steam' from a boiling kettle or steam-engine.

STEAM FOG, produced in the same way by sufficiently cold air above warm water; and

SMOKE FOG, which is exactly what it is called, being most troublesome where air or ground traffic and heavy industry are equally intensive and too closely crowded.

5. THUNDERSTORMS

15 September has an afternoon thunderstorm showing familiar features, including wind gusting most strongly with the first heavy rain and then falling so nearly calm as to let a slight mist form in the wet air.

The electrical working of a thunderstorm is not yet fully understood. Many years ago it was suggested that the bigger a water drop is, the faster it falls, so the faster is the up-current required to support it; but it cannot exceed a certain size without breaking up; while if it is broken by contact with anything (air, for example), both of them acquire electric charges (equal but opposite), so that any up-current more than strong enough to hold up the biggest water drops must result in this kind of separation of electric charges, which by continued growth of water drops in a cloud must be multiplied into great charges causing such electrical tension as may be released by discharges many miles long. Atmospheric electricity is a big subject, but at least it is fairly certain that a thunderstorm depends largely on sufficiently vigorous *convection* (with clouds).

This, in turn, as we already know, occurs with any sufficiently high w.p.t. of rising air when there is steep (super-adiabatic, wet or dry as the case may be) temperature lapse rate of upper-air environment. Warm wet air at the earth's surface, for instance, with cold upper air, such as polar air over the sea in the winter or else over hot land in summer, especially in the tropics—those are ideal for thunderstorms. The tropics will best illustrate the idea. Owing to the heat of luxuriant wet vegetation inland, and of neighbouring seas, the lower air has the world's highest moisture content. So it is full of latent heat. This heat is revealed set free in the clouds, which thus mark effectively hot upper air.

Among our first principles in the first chapter you may remember that winds blow all the time into the tropics over the ground, not from due north or south but from a more easterly quarter. Those are the so-called trade winds. Where do they meet and rise? Where it is effectively hottest. So at any particular height above the tropics the meeting-place of the ex-northern and ex-southern trade winds is found wherever the *clouds* all extend up to this height. The meeting place at a *great* height is therefore found where the clouds all rise to a great height, for which they need th greatest possible w.p.t. So it is not necessarily where the tropics are hottes in the dry sense, but only where heat and moisture compromise to give the highest wet temperature. Over open sea this is never far from the thermal equator (the warmest sea zone); for wherever the sea is warmest the air which touches it must be warmest and (*if given time enough*) will have had its chance to pick up the most moisture. Only if one of the trade winds is slower than its opponent will it have had more time and so become damper, displacing the upper-air meeting place some distance away from the thermal equator.

This generally cloudy or even stormy meeting place of the trade winds aloft over open sea in the tropics is the notorious DOLDRUMS zone. Its regular changes are therefore those of the thermal equator, which reckons to lag about two months behind the sun in its annual movements north and south. Its *irregular* changes, of which unfortunately little has ever been said among old traditions, are partly those of the trade winds and thus of the subtropical high-pressure belt out of which they come. Again to quote first principles from our first chapter, the polar front itself finally 'surging into the tropics...feeds the subtropical high-pressure zone with its fresh high pressure so that it may make its influence felt all the

	Place no.	Date 1936	Time G.M.T.	Weather letters	Cloud T.	Cloud L.	Cloud M.	Cloud H.	Vis. V.	Wind D.	Wind F.	Temp. F.	Temp. C.
	01	15.9	0700	o m$_0$	8	5	–	–	5	NE	1	55·0	13
	01	15.9	0930	bc m$_0$	5	5	7	–	5	NE	3	55·0	13
	01	15.9	1340	c	7	o	7	–	6	NE'N	3	59·9	15
[a]	01	15.9	1410	bc	4	3	8	6	6	NE'N	2		
[b]	01	15.9	1445	bc U	5	3	8	4	7	NE'N	2		
	01	15.9	1448	c U+t+pr	7	3	–	–	7	NE'N	2		
[c]	01	15.9	1450	c U T l+pr	7	3	–	–	7	NE'N	3		
[d]	01	15.9	1454	c t l PR	7	3	–	–	7	NE'N	5		
	01	15.9	1510	c t l r	7	3	–	–	6	NE'N	4		
	01	15.9	1520	c t l r	7	3	–	–	6	NE'E	2		
	01	15.9	1535	c r$_0$	7	3	–	–	6	NE'E	1		
[e]	01	15.9	1605	bc m$_0$	5	3	6	3	6	NE	1		
	01	15.9	1635	bc m	4	3	o	8	4	Calm			
	01	15.9	1830	b	1	5	o	o	6	NE	1		
	01	15.9	2100	c m$_0$	6	5	o	o	5	NE	1		
	01	16.9	0650	c f	6	o	3	–	2	NNE	1	53·3	12
	01	16.9	0800	c f	7	o	3	–	3	NNE	2		
	01	16.9	0900	bc m	3	1	3	–	4	NNE	2	57·2	14
	01	16.9	1000	c	7	3	3	–	6	NE'N	3	59·3	15
	01	16.9	1030	bc	4	3	6	–	6	NE	3	61·2	16
	01	16.9	1310	c pr	7	9	–	–	7	NE'E	3	61·5	16
[f]	01	16.9	1340	c	7	3	1	–	7	NE'E	3		
	01	16.9	1615	c	7	3	1	–	6	NE'E	2	63·0	17

[a] Solar halo.
[b] Black sky to north.
[c] Lightning increases as storm approaches.
[d] Lightning flashes single, not multiple.
[e] Valley mist.
[f] Medium clouds moving from SE.

	Place no.	Date 1936	Time G.M.T.	Weather letters	Cloud T.	L.	M.	H.	Vis. V.	Wind D.	F.	Temp. F.	C.
	01	16.9	1830	o	8	5	2	–	6	NE'E	3	60·3	16
	01	16.9	1950	bc	5	o	5	–	6	NE'E	3	59·4	15
	01	17.9	0610	o m	8	5	–	–	4	NNE	2	58·0	14
[g]	01	17.9	1000	c m$_0$	7	2	7	–	5	NNE	2		
	01	17.9	1210	c pr$_0$	7	3	7	–	6	NNE	2		
	01	17.9	1500	c	7	3	7	–	6	NNE	2	66·1	19
	01	17.9	1730	o pr$_0$	8	3	2	–	6	NNE	2	63·4	18
	01	18.9	0700	o m$_0$	8	o	2	–	5	—	1	58·0	14
[h]	01	18.9	0900	c m	7	o	5	–	4	—	1	62·1	17
[i]	01	18.9	0925	c m$_0$	6	3	6	–	5	—	1	61·1	16
[j]	01	18.9	0930	c m$_0$	6	3	6	–	5	—	1		
[k]	01	18.9	0935	c	6	3	–	–	6	NW'W	1		
[l]	01	18.9	0940	c	7	3	–	–	6	—	1	61·8	17
	01	18.9	1530	bc	5	3	6	–	6	WSW	2		
	01	18.9	1730	bc	5	3	6	–	6	WSW	1		
	01	18.9	1820	bc m$_0$	4	5	6	o	5	WSW	1	62·1	17
	01	18.9	2100	b m$_0$	o	o	o	o	5	WSW	1	60·0	16
	01	19.9	0630	o f	8	5	–	–	3	Calm		56·3	13
[m]	01	19.9	0900	c f	7	5	–	–	2	NE	2	56·5	14
	01	19.9	1530	b f	1	o	o	1	3	NE	2		
	01	19.9	1600	bc f	3	o	o	1	3	NE	2		
	01	19.9	2040	b m	o	o	o	o	4	NE	2	57·3	14

[g] Medium clouds moving from E. Rather gloomy sky. Rain for 20 min.
[h] Medium clouds moving from SW.
[i] Low cloud base moving from E.
[j] Relative motion higher up from WSW.
[k] Cloud top moving from WNW.
[l] Smoke-drift directions conflict.
[m] Water-fog layer lifting and breaking, leaving smoke haze.

way to the equator. Meanwhile, of course, being strongly heated over the tropics it becomes tropical air and rises well laden with moisture over tropical seas.' In fact it becomes EQUATORIAL air. Like polar air round a depression, it rises unhindered except by its own viscosity or internal friction. But convection illustrates survival of the fittest. The air flows in from each side of the doldrums zone, out with high clouds at the top, and *down again on each side* to suppress all its weaker neighbouring trade-wind cumulus clouds, which are therefore kept small, unable to break through their 'ceiling' till near the doldrums. Not that this is the whole story, or even the only version of this part of the story. But it conveys very broadly what happens.

Meanwhile the height of *base* of these clouds depends on how dry the ground is, how low the surface r.h. is. Under the subtropical high-pressure belts of fine weather all the year round out of reach of the rainy doldrums, the land is desert so dry that no clouds can normally be formed at all.

A polar or cold front, however, raises the 'ceiling' and so may allow clouds to form underneath it. As the ceiling rises towards the doldrums anyway, these clouds grow abnormally tall and finally give the world's biggest thunderstorms. At first they are only abnormal in electrical tension (owing to steep lapse rate of environment temperature) whilst still surrounded by very dry air that tends to dissolve them; but later it is in sheer size and endurance that they become great. That is how a cold front is suspected of making its influence felt all the way to the equator.

The tropics have thus shown a general principle of thunderstorms. Continental air in the summer, particularly under high pressure, normally cloudless though tending to thunder after the heat of each day, is made abnormally thundery by any cold front, however old and hitherto innocuous. If, moreover, the front becomes quasi-stationary, then it is rather like the equatorial doldrums, a *rainy* zone flanked by *thundery, fair and fine* weather zones, according to distance out from the frontal low-pressure trough, as may be traced on a map.

The 'business end' of a storm cloud is fairly high up, and therefore moves not with the low-level wind (which, in fact, naturally has some component *into* the cloud base anyway) but with the upper winds. So a storm familiarly comes up 'into the wind', this being merely the low-level wind. Exceptionally it may appear to move against the upper winds, but then it is rather being *fed* by conditions below it which freshly develop it up-wind aloft whilst leaving its previous centre to die down. Recent researches, moreover, have made it clear that even a single storm may have more than one great upward current or 'centre'.

16–18 September, though not actually thundery, show much the same tendency as 15 September. Morning fog on **19 September** is dried at midday but leaves thick smoke haze owing presumably to the light drift of unstirred stable air from London.

To conclude this section on thunderstorms you may be interested in ways of observing them. First is the ordinary Meteorological Office synoptic weather reporting system in which paid observers either telephone, telegraph, teleprint or otherwise promptly transmit weather reports at regular hours to their central office. Whether of thunder or anything else, their reports are not merely kept on record but are used at once for charts and forecasts.

Thundery weather, of course, can also be picked up by radio. Special direction finders at various widely spaced stations are used to locate distant sources of atmospherics, which are likewise reported at once to a central office for forecasters' use, while actual cumulonimbus or even cumulus clouds when nearer, even if unseen by eye, can nevertheless be 'seen' by RADAR.

For the purpose of more detailed records, not for immediate but for later research, there is in this country a Thunderstorm Census Organisation (conducted by S. Morris Bower, of Langley Terrace, Oakes, Huddersfield, Yorkshire, England), in which anyone is welcome to take part, requiring just a brief though careful report (usually on a specially printed postcard form, of which a stock is always supplied on request) of any thundery weather observed.

'Any data, however incomplete', say the organisation's instructions, 'are worth sending. Copies of local press reports form a welcome supplement to the records.'

'Evidence', too, 'that no thunder or lightning has been observed in an area since the last communication, or during another stated period', is very useful.

'In estimating distance the popular belief that the storm is one mile distant for each second between the lightning and the thunder is quite erroneous. Actually one mile is indicated by FIVE SECONDS.'

For more immediate research, too, records are appreciated as soon as possible after a storm. Maps of notable storms are then issued.

V

AIR TEMPERATURES

1. TEPHIGRAMS

Now that we have been introduced to upper-air temperatures (u.a.t.) and how they affect the weather with cumulus clouds and thunderstorms and other kinds of convection, we can deduce how to chart them.

'The appropriate chart', you remember among our first principles in Chapter I, 'is ideally a map of the weather at all levels, one time and one place. That is an aerological or upper-air chart.' Which weather elements is it to show? Simply our ABC weather elements, temperature, pressure and moisture, which are completely determined by any *three* of the following, taking two from one column and then one from the other:

Pressure	Pressure
Temperature	Wet temperature
Potential temperature (d.p.t.)	Wet potential temperature (w.p.t.)
Saturation moisture content (s.m.c.)	Actual moisture content (m.c.)

A chart is to show them by lengths or distances. As there are three, a solid model could show them all together, say by latitude, longitude and altitude, which determine a *point*. Their distribution will then appear as a set of points, a line or *curve*. On paper, however, having length and breadth without depth, only two at a time can be shown by points or a curve. Pressure and temperature give a 'u.a.t. curve', while pressure and *wet* temperature give a 'wet u.a.t. curve'. As the two curves should suffice to tell the whole story, it only remains to learn how to read them.

Remember these rules:

1. Our basic unit of pressure is called a BAR. In popular units it is just under 1 ton per sq. ft.
2. To avoid, however, not only decimal points but also other meanings of the word 'bar' we take as our practical unit not the bar itself but the MILLIBAR (abbreviated to 'mb.'), that is a thousandth part of a bar.
3. *Approximately*, then, the pressure of the atmosphere is at zero-level (ground- or sea-level) 1000 mb.; at a height of 10,000 ft., 2 miles or

56

3 km., 700 mb.; and at a height of 30,000 ft., 6 miles or 9 km., 300 mb. Those may be taken as our bottom, middle and top standard levels. Or as the top, instead, we might take 500 mb.

4. Rising dry air cools 10° C. per km., say from 30° C. at the ground to 0° C. at the middle level (3 km.), and to −30° C. at 6 km.

So if *temperature* is marked uniformly *across* the page ('west' to 'east'), while *height* is marked *up* the page ('south' to 'north'), then the dry rising air is shown by a line that slants across from bottom right ('south-east') to top left ('north-west'). As its slope or direction across the page must represent the dry adiabatic lapse rate of 10° C. per km., which, of course, is the same for any dry air whatever, you see that the u.a.t. line for any adiabatically ascending or descending dry air must lie 'south-east to north-west'. As the temperature where it comes to the bottom is already defined as the air's d.p.t., these parallel 'south-east to north-west' lines must be *potential temperature lines*. You can add them to the original network of 'west-east' lines and 'south-north' lines in rather the same way as you can add magnetic compass lines on to an ordinary map. In each case you have three variables, of which any two will determine the third.

You may also remember that any two of these three will suffice to determine the air's *saturation* moisture content, so we can add s.m.c. lines too. Thus we have shown, you see, all the four quantities—pressure, temperature, potential temperature (d.p.t.) and saturation moisture content (s.m.c.)—which are listed in the first of our two columns opposite. In such a chart, or AEROGRAM, the standard lines form a kind of grid or network, on which the actual u.a.t. are shown by a curve through a set of points. If, for example, the upper air is so stirred as to have uniform d.p.t., then its u.a.t. 'curve' is simply a 'south-east to north-west' d.p.t. line. That represents temperature lapse rate 10° C. per km. while *potential* temperature lapse rate is zero. So we show an ADIABATIC atmosphere. With lapse rate any greater than this, shown therefore by a *more nearly 'east-west' u.a.t. curve*, being potentially warmer at the bottom than at the top, the air tries to overturn, with convection. That is UNSTABLE air. If, on the other hand, the upper air has not uniform potential temperature but uniform ordinary (dry) temperature instead, then its u.a.t. 'curve' is simply an ordinary temperature line up the page. That represents temperature lapse rate zero while d.p.t. lapse rate is −10° C. per km., as the air potentially warms with height instead of potentially cooling with height. So we show an ISOTHERMAL atmosphere. Most air is intermediate between isothermal and adiabatic.

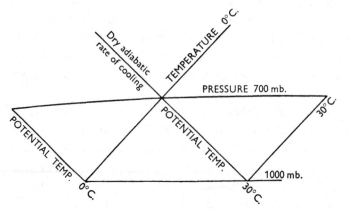

Fig. 6. Tephigram foundations.

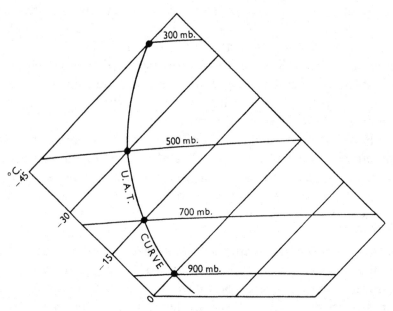

Fig. 7. Illustration of u.a.t.

Polar air

Pressure in mb.	Temp. °C.
900	0
700	−15
500	−30
300	−45

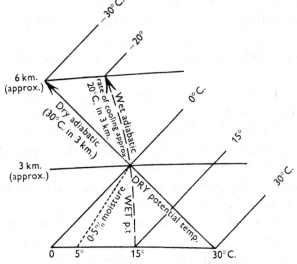

Fig. 8. Tephigram construction.

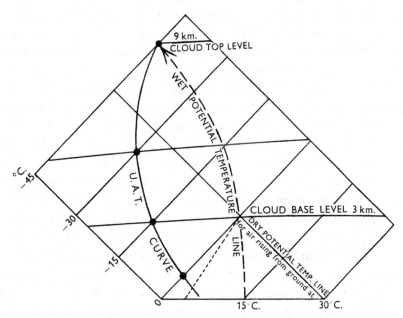

Fig. 9. Illustration of polar air cloud formation
from 3 to 9 km. over hot ground.

59

Just as maps can be drawn on different projections, different in shape yet all showing every place in its right latitude and longitude, so may aerograms differ in form. Our favourite one is the TEPHIGRAM, where with zero level along the bottom we have temperature lines not 'south to north' up the page but 'south-west to north-east' at right angles to the 'south-east to north-west' d.p.t. lines.

Now we have to take moisture into account. Suppose the air at the bottom has temperature 30° C. but m.c. only about ½%. Then its dew-point is about 5° C. (about 40° F.). Just as the figure of 30° C. at the bottom level (1000 mb.) provides a fix on the air's d.p.t. line, so the figure of 5° C. here provides a fix on its moisture line. As the air rises with d.p.t. constant at 30° C., its ordinary temperature falls to 0° C. at the middle level (3 km. or 700 mb.). Meanwhile its dew-point has also fallen to 0° C. So its ordinary temperature and dew-point are now the same, which means it is saturated. Here on the chart is the meeting point of d.p.t. line with moisture line, defined as the NORMAND POINT for that particular air. Here, too, in the sky itself, we might expect to find a cloud base.

Wet and dry temperatures now being the same, all further cooling of the rising air must be at the *wet* adiabatic rate of about 20° C. per 3 km. instead of the dry rate of 30° per 3 km. So the u.a.t. curve for the rising air must lie in between the 'south-west to north-east' (isothermal) temperature line of 0° C. and the 'south-east to north-west' (dry adiabatic) d.p.t. line of 30° C. The higher it goes, in fact, the less moisture can the air hold, so the less difference is there between wet and dry rates, and so the more nearly 'south-east to north-west' does our curve lie. In other words it leans over 'westward'. As we have worked out (in the last chapter) that the *wet* temperature must have been falling at *wet adiabatic* rate from the very beginning, while our curve shows this wet rate in the top half of the picture, we find the original or potential wet temperature (our old friend w.p.t.) simply by extending our curve right down to the bottom (1000 mb.). Here we find it at about 15° C. (about 60° F.). Likewise other w.p.t. lines can be drawn up from the bottom with their tops all leaning towards the left or 'west' side of the picture as if they were blades of grass in a breeze.

As dry air is unstable wherever its u.a.t. curve lies more nearly 'east to west' than the 'south-east to north-west' dry adiabatic direction, so is saturated air unstable wherever its u.a.t. curve leans more to the 'west' than these grass-blade wet adiabatics. Then it tries to overturn and form masses of cloud such as cumulus and cumulonimbus. As the air is saturated, this u.a.t. curve must be the same as the *wet* u.a.t. curve, which in turn cannot have been altered by any evaporation of raindrops falling into the air from above. Saturation of air by falling rain must therefore release instability wherever the wet u.a.t. curve leans more than wet adiabatic.

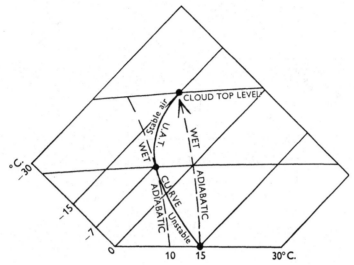

Fig. 10. Illustration of saturated air u.a.t. compared with wet adiabatic, showing wet air stability.

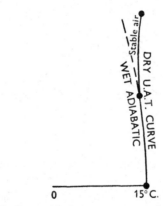

Fig. 11. Ordinary (dry) u.a.t. compared with wet adiabatic, showing stability with no clouds at w.p.t. 15° C.

That is wherever the air's w.p.t. decreases with height, being less at the top than the bottom, so that it is effectively cooler above than below, even though not potentially so in the *dry* sense.

On the other hand, the air may be saturated not by raindrop evaporation conserving w.p.t. at each level, but by air-mass ascent over hills or fronts. Before saturation, you will remember, this process leaves not only w.p.t.

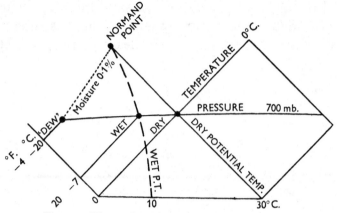

Fig. 12. Illustration of wet, dry and dew-points.

Pressure (mb.)	Temperatures (° C.)		Dew- or frost-point
	Dry	Wet	
700	0	−7 (ordinary)	−20
	30	10 (potential)	

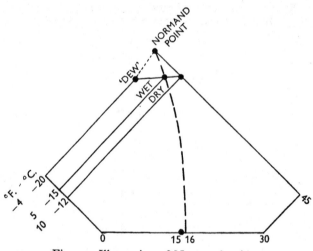

Fig. 13. Illustration of Normand points.

Pressure (mb.)	Temperatures (° C.)		Dew- or frost-point
	Dry	Wet	
500	−12	−15 (ordinary)	−20
	45	16 (potential)	
1000	15	15	

(Dry, wet, dew and Normand points all the same)

but also d.p.t. and m.c. unaltered—though not at each level but rather *in each piece or element of rising air*, which is quite a different matter. As the Normand point for each element of rising air, being the meeting point of its d.p.t. and m.c. lines upon the chart, must thus stay fixed till the air is saturated, you see that the NORMAND CURVE which links the Normand points together must be the same originally as finally when it becomes the actual u.a.t. curve. But then any instability will as usual be shown wherever the curve leans more to the left of the picture than wet adiabatic. *So the Normand curve of dry air, if leaning more than wet adiabatic*, shows an otherwise hidden instability that can be released by air mass ascent.

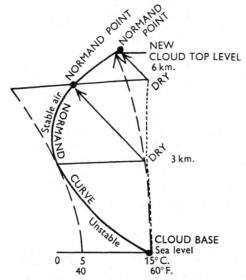

Fig. 14. Normand-curve u.a.t. produced by air-mass lifting, showing instability with clouds up to 3–6 km.

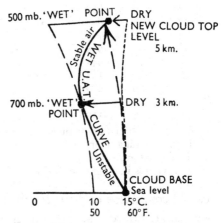

Fig. 15. Wet u.a.t. produced by evaporation cooling the air, showing instability with clouds up to 3–5 km.

2. POTENTIAL INSTABILITY

20 September shows signs of depression developing from the south with a front coming over our easterly air stream. Medium clouds increase while cumulus clouds decrease rather as on 6 September. The irregular character of the heavy rain suggests upper-air instability. That is confirmed by thunderstorm. But it is certainly not after normal heat of the day, so it must be a frontal thunderstorm. In fact it must be the kind of instability that is hidden until set free by air-mass lifting, when the air's Normand curve leans more to the left of a tephigram than wet adiabatic lines. That

	Place no.	Date 1936	Time G.M.T.	Weather letters	Cloud				Vis.	Wind		Temp.	
					T.	L.	M.	H.	V.	D.	F.	F.	C.
	01	20.9	0820	c	7	8	–	–	6	ENE	4	60·1	16
	01	20.9	0910	b	2	1	5	0	6	E'N	5		
	18	20.9	1130	bc	5	0	5	–	6	E'N	6		
	18	20.9	1200	bc	3	1	5	0	6	E'N	5		
[a]	18	20.9	1545	c	7	0	7	–	6	E'N	5		
	01	20.9	1620	o rr	8	7	2	–	6	E'N	4		
[b]	01	20.9	1645	o R	8	7	–	–	6	E'N	4		
	01	20.9	1850	o	8	7	–	–	6	E'N	4	62·0	17
	01	20.9	2030	o r	8	7	–	–	6	E'N	4		
[c]	01	20.9	2115	o t l R	8	7	–	–	6	E'N	4		
	01	21.9	0630	c	7	7	7	–	6	SW	3		
	01	21.9	0930	c	7	5	7	–	7	SW	3	63·1	17
	01	21.9	1110	o rr/id$_0$	8	7	–	–	7	SW	3		
	01	21.9	1150	o id$_0$	8	7	7	–	7	SW	1	64·8	18
	01	21.9	1530	c	7	5	7	–	6	WSW	1	64·8	18
	01	21.9	1630	o r$_0$	8	5	2	–	6	W'N	2	64·0	18
	01	21.9	1650	o/r	8	5	–	–	5	WNW	2	63·9	18
	01	21.9	1800	c m$_0$	7	5	–	–	5	NW	1	62·0	17
	01	22.9	0700	o F	8	5	–	–	1	Calm		57·2	14
	18	22.9	0935	o f	8	5	–	–	2	Calm			
	18	22.9	0945	o m	8	5	–	–	4	Calm			
	18	22.9	1135	b m$_0$	1	1	0	0	5	SW	1		
	01	22.9	1400	bc	3	2	0	0	6	SW	1	67·8	20
[d]	01	22.9	1530	c	7	3	0	0	6	—	1	67·2	20
	01	22.9	2110	c m	7	–	–	–	4	Calm		63·0	17

[a] Thick altostratus predominating.
[b] Rain becoming temporarily heavy, and thereafter intermittently heavy until 1730.
[c] Thunderstorm.
[d] Low clouds moving from SSW.

is called *convective* or *potential* instability. Generally, of course, it implies a wet u.a.t. curve also leaning more to the left of a tephigram than wet adiabatic lines, meaning a DECREASE OF W.P.T. WITH HEIGHT, when instability is equally well set free by saturation with raindrops. Falling raindrops, you see, are effectively *wet bulbs*, so that they tend to take up the air's wet-bulb temperatures but lag behind them when falling from colder air. Although they cannot then quite saturate the air, sufficient stirring will finish the job by cooling the top as well as warming the bottom of a yet stable layer of air. As soon as any clouds (TURBULENCE CLOUDS such as stratus or ragged low clouds of bad weather, C_L type 7) are thus formed at the top of this stirred layer, they find themselves effectively cooler above than below, and so overturn to form more.

Any kind of decrease with height in the air is known as LAPSE. All we need for potential instability is therefore a lapse of w.p.t. As it implies also lapses of d.p.t. and/or m.c., which in turn are naturally produced by convection on warm land or sea, it is most likely over this country in upper southerly winds that at the bottom have been made first very moist in the Mediterranean Sea and then very warm over France, particularly in the early summer when the air higher up has not had time to pick up anything like as much heat as the air underneath it.

21 September, with a south-west wind, seems to find us just behind the centre of the depression which was suspected yesterday. The weather looks frontal but not distinctly warm-front nor cold-front in type. There is just a lot of dense cloud and rather persistent though not necessarily continuous rain. This sort of weather is apt to be caused by the general convergence of air into the centre of a depression, to which we have been introduced in connection with wind.

To-day, however, there is little wind, so that after the rain there is mist which is probably helped by London smoke haze. Compare 9 and 14 September which both show sea air fog-point near the sea temperature of about 15° C. That is easily reached at night-time now, after day maximum temperature only 18° C. So come the first autumn mists.

22 September begins accordingly with fog, which lifts and finally breaks into scattered low clouds when enough heat gets through. Temperature reaches 20° C., as is appropriate to fine calm days at the equinox in these latitudes when the sun's midday height above the horizon is about 36°. It is the same at midwinter, too, at the edge of the tropics (latitude 30°) where day maximum temperature of 20° C. and night minimum temperature 5–10° C. and average temperature about 15° C. are likewise largely determined by the height of the sun in the sky.

So we are introduced to the subject of the sun's radiation of heat, which will now be described as simply as possible. Avoiding the technical language of radiation physics we shall use units that make the numbers so simple that we can work them out in our heads.

3. RADIATION

The best things in life are free. Sunshine is free, and though you may live in darkness or only in artificial light you still cannot live without certain natural things that depend on sunshine. The air also depends on sunshine. Meteorology might, indeed, begin with the simple fact that through any area facing the sun at right angles, anywhere near the earth (but outside the atmosphere), there is an almost constant flow of just over a million calories per square metre per hour.

A halfpenny held up at arm's length is more than enough to eclipse the sun. How far away can you take it, before it fails to cover the disc of the sun? Try it. Try to be accurate, too. You might even compare the midday sun with the setting sun. You will be surprised when you find how far it can be taken. The halfpenny, which is an inch wide, can be taken 9 or 10 ft., just over 110 in. The sun's distance is therefore just over 110 times its own diameter. So it is just over 220 times its radius. A beam of sunshine coming straight out from one centimetre of the sun's surface must thus cover just over 2·2 m. here at the Earth. Now a 2·2 m. square has an area of about 5 sq. m., thus receiving over 5 million calories every hour. Nearly 6 million, in fact. All must have come from a single sq. cm. of sun. So if the sunshine power unit be 1 cal. per sq. cm. per hr., which is about 1 watt per sq. ft., then the sun's output must be nearly 6 million units.

How can we deduce its temperature? You know that our temperature degrees are defined by the step between normal freezing- and boiling-points of water, of which for example a Centigrade degree is just 1%. Suppose we take the whole step as our unit instead. The temperature of our tropical seas is then almost exactly three of these steps above absolute zero, while that of your body is 3·10. Water boils at 3·73 and freezes at 2·73.

With these particular units suppose we *square* the temperature (the square of 3 steps, for instance, being 9), and then square again ($9^2 = 81$). Then for a perfect radiator the product is almost exactly *twice the radiation power*.

With more orthodox units, of course, it would not be just *twice* but would be some other (less simple) multiple of the power. That is the well-known physical constant factor named after Stefan. We have simply chosen our units to make easy figures.

Our tropical seas, for instance, with temperature 3 units, radiate just one half of 81 cal. per sq. cm. per hr. That is about 40. For the sun, on the other hand, the figure of nearly twice 6 million is nearly the square of 3600, which in turn is the square of 60. So the sun's effective surface temperature must be nearly 6000° C.

To see how the earth must balance its radiation we might well think first of the moon, which is not complicated by air or water or life to use up any heat, yet is just about as far from the sun as we. All we can say at first is that at its equator, with sun overhead, each square metre of ground every hour must receive nearly 1·2 million calories. That is nearly 120 of our units. How much is absorbed is not easy to say, but obviously the square of the square of its temperature will be getting on for nearly twice 120.

Twice 120 is 240. Well, 256 is the square of the square of 4. So the hottest part of the moon should be getting on for 400° C. above absolute zero, like boiling water.

What about the average figure? What is the average sunshine between sunrise and sunset? With the sun only 30° above the horizon, the beam of its rays which would put nearly 120 units of power through an area facing them, broadside on, must now fall at an angle so as to cover twice as much area of horizontal ground, so that each unit area of the ground can only receive half as much. At sunset, of course, the fraction has fallen to zero. The mathematical average of this varying fraction is about $\frac{2}{3}$. So the average rate of absorption of sunshine by the ground is about two-thirds of 120, which is 80 units. That is for the period from sunrise to sunset. For the whole period from one sunrise to the next sunrise the average is only half as much, the time being doubled without any further sunshine. That makes 40. Twice 40 is almost the square of 9 which in turn is the square of 3. So this average equatorial temperature of the moon should be much the same as the earth's which is 3 steps or 300° C. above absolute zero (−273° C.).

The words 'this average', by the way, have just been used with a double meaning on purpose to remind you that the process of averaging may be done in different ways, giving different answers. The radiation actually coming off the moon, moreover, has often been measured.

Now we come down to earth. On the earth's equator with midday sun overhead as it was in the case of the moon, heat comes from the sun at the rate of nearly 120 units. Generally at least 10% is reflected or scattere back by the atmosphere, and anything up to 80% of the rest is reflected or scattered by the earth's surface itself. But apart from rays at low angles only a little will come off the sea, while most equatorial lands are dark enough (mostly with jungle) not to reflect more than 20%. So in clear skies at the equator nearly 90 units may be absorbed. From the sea or from land at an equal temperature we already know that 40 units come straight out again in the form of radiant heat. As these cannot balance nearly 90 it ought to get hotter. But there is an atmosphere to share this surplus. How can it do so? We know that the air on the ground, whenever sufficiently wet or hot and thus buoyant, must rise. This stirring process of convection will try to go on until enough heat has been taken up to give all the air, as far as possible, the same potential temperature, wet or dry as the case may be. How far is possible? How far down and how far up above ground? It is down to just a few metres or nearly enough right down to just a few feet above ground. That is about the height of ourselves or of our thermometers. Above it convection goes up to the levels beyond which the air all the time has been potentially warmer (again in the wet or dry sense as the case may be). So it depends on how far the old upper air has already been warmed otherwise. In the tropics, for instance, having come all the way from cooler regions, it cannot have already been warmed enough to stop convection very low down, so convection there may go up high.

'Ground', up to now, has meant land or sea. The time has come for us to distinguish them. Convection over the sea, being naturally of wet air with clouds, takes heat mostly from evaporation of water. A few millimetres of water each day, less than a gram over each sq. cm. of sea, may take up hundreds of calories every day. Averaging up to something like 15 per hr., it should thus enable convection altogether to take up to 20 or 30 units of power. Sunshine meanwhile goes deep down through the sea. As its heat is thus not all used up in a shallow layer but spread instead through much water, it hardly alters the temperature at all. Tropical waters stay at about 300° on the Centigrade scale above absolute zero, so that they go on steadily radiating upwards with a power of 40. There again, however, the air joins in. Whilst nearly transparent to the sun's direct radiation as *light*, the air's invisible water vapour is partly opaque to the earth's radiation of heat. It duly takes in whatever comes up from the earth, sends some out all the time according to its own abundance, distribution and temperature, and lets the rest through. Its absolute temperature is between 200° and 300°. If equivalent, therefore, to some perfect radiator at 250°, it should send out half of the square of the square of 2½ (steps). That is about 20 units. That is what the earth would get back from its own radiation of 40.

Actually the effects of the air's water vapour and other gases are only known roughly. At any rate they are complicated to estimate. But at most, so long as the air keeps its water in the form of invisible vapour rather than clouds, it can hardly send down to the earth more than 30 units. Very dry air can send very little. So the earth's balance of lost radiation must be at least 10. Its minimum, in fact, occurs under the dampest air, namely tropical sea air already filled (by evaporation or convection) with anything up to 20 units. So we account altogether for 30. Since not more than three-quarters as much sunshine can be absorbed here as was on the moon, you see that the average income between one sunrise and the next sunrise cannot be more than three-quarters of the moon's equatorial figure of 40. So our 30 units are balanced.

Dry land, on the other hand, absorbing all in a shallow layer, can warm up and thereby radiate more to balance the incoming sunshine power. At highest it reaches 330–340° (about 60° C.), at which it radiates 60 units. Air sending back 20 would then be letting out 40 while convection with d.p.t. up to 320° (nearly 50° C.) from just a few metres above the ground must be taking up all the rest. A few hours later in the day, when convection has ceased and the sun at an angle of less than 30° above the horizon cannot now be feeding the ground with more than about 40 units, the temperature will have been able to start to fall. Radiation, although at first perhaps 40, will therefore only average 30, say, for the next 12 hr. 30, multiplied by 12, is 360. The units are cal. per sq. cm. Suppose that the cooling layer of dry ground has heat capacity equivalent to 10 cm. of water. Then it would cool 36° C. in the night. Thus, starting from 60° C., it would reach 24° C. The air's shade temperature would follow it down to about 25–30° C., say 300° absolute (300° K.).

Now having inferred the day's highest and lowest air temperatures we can estimate roughly the heat of the next day's dry convection. First of all in the morning as all the previous day's uncooled hot air lies above, convection must be confined to a shallow air layer having average depth, say, $\frac{1}{2}$ km., about 50 gm. per sq. cm., having heat capacity roughly equivalent to 10 cm. of water. Whilst its top is unchanged, the bottom is heated some 10° C. (300–310°). Taking the average then as 5°, we infer 50 calories taken up. Later convection is several times as deep. With an equal rise of temperature but four times the depth of air, for example, we should infer not 50 but 200 calories taken up. Total 250 in 8 hr. will then represent an average power of just over 30 units. Averaged over the whole 24 hr. day we can call it 10. It may be more. Anyway it is comparable with convection at sea.

Notice then the day's average temperature of the air over land. It is 310° absolute. That is greater than over the sea. Just as the *actual* temperature cannot fall from its greatest figure until long enough after noon for the sunshine decreasing to let it, neither can the daily *mean* value fall from its 'summer' figure till long enough after midsummer for the daily average sunshine decreasing to allow it. Thus, after the yearly midsummer solstice (about 21 June in northern latitudes), comes the 'continental midsummer' of temperature and climate. Generally the drier the land the shorter the lag. We might, for instance, take 1 January as northern continental midwinter, 1 July its midsummer, and therefore 1 April and 1 October as the ends of the summer or winter half-year.

Having seen tropical continental midsummer conditions we must think of midwinter. Taking the minimum as at the edge of the tropics, say in latitude 30° where the sun is not more than 36° up, and shines for less than 12 hr., the daily average sunshine rate can only be half its midsummer figure of 30 units. How does the ground balance that? Averaging about 290° in absolute temperature it duly radiates just over 30 units. Air then returning 20 must let out 10. If convection also at only half its midsummer averate rate thus adds 5, the 15 are balanced.

At midday the ground may reach 300° to radiate 40. Air still returning 20 must now let out 20. Then with convection at the same time taking up 20 it can balance the 40 coming in from the sun. Again within a few hours all convection has ceased, and with sunshine less than 20 the ground can cool down. Averaging 280–290° it will radiate 30 units, of which as usual the air may send 20 back, so it loses 10 cal. per sq. cm. per hr. The winter air will actually be rather drier and thus send back less than 20. If it only sends 10, for instance, it lets the ground lose 20. So with a 'water equivalent' of say 10 cm. the ground will cool through

$$\frac{12 \times 20}{10} = 24° \text{ in 12 hr.}$$

That takes it down from 300° to 276°, which pulls the air temperature down to about 280°. Then, heated during the morning first from 280° to 288° with convection as before to $\frac{1}{2}$ km. with the same water equivalent for heat, the air will take up 40 units. In the summer with a 10° temperature rise we reckoned 50 units, so now with an 8° rise we should reckon 40. Afterwards heated over the ground from 288° to 293° (20° C.) with convection say *three* times as deep, it must take up 75. Altogether in 6–8 hr. we see about 120 used up. That is an average of 15–20. Averaged over 24 hr. it is of course only 5.

Between the daily mean shade temperatures of 310° (midsummer) and 290° (midwinter), you see the same yearly average figure 300° for the dry land as for the sea. At the edge of the tropics, too, 310° and 290° would be about the limits of *actual* temperature on any one day about midway between midsummer and midwinter, namely some time near the first day of October or April. At some place nearer to the equator at that same time it would of course be hotter, so that the same range (310–290°) would only be experienced there on days nearer to midwinter. Even closer to the equator the dry air is left behind, and finally so is all really dry land. The damp conditions prevent the temperature ranging as far as 20° at all. On many equatorial coasts, for instance, it stays between 295° and 305° in damp air all the time. The dry air at 310°, with moisture content only $\frac{1}{2}$% of the air, feels cooler than the damp air at 305°, which is a full 2% water vapour. In it, you see, the evaporation of anything wet can be sufficient to cool it down to 290°, whereas in the damp air the corresponding wet temperature cannot be less than 300°. It is those wet temperatures that make all the difference.

Meanwhile what happens at home in this country? Suppose we count it as continental, with latitude 50–60°. Our ideas should then hold good for many other such places. At midsummer, with sun up to 60° above the horizon, as at many places inside the tropics in winter, or as at the edge of the tropics in autumn or spring, the air temperature may range from 290° to 310°, with a day's average of 300°. At the ends of the summer half-year near early October or April, it must be much the same as we found at the edge of the tropics for midwinter when the sun's midday height in the sky was about the same. The figures then, you remember, are (in round numbers):

Air maximum temperature 20° C.; minimum 5–10°; average near 15° C.;
Ground maximum 20–30° C.; minimum 0–10°.

Those are easy enough to remember. Now what about midwinter? In fine (continental) weather in winter the ground takes in little or no sunshine but just goes on sending out 30 units (at 280°) or anything down to 20 (at 250°), of which the air must be sending back considerably less and thus letting some of it out all the time.

So the ground just goes on losing heat at any rate up to something like 10 cal. per sq. cm. per hr. With the old equivalent of 10 cm. of water, it would then cool at any rate up to about 1° C. per hr., unless somehow fed with heat from elsewhere. Wind turbulence up to $\frac{1}{2}$ km. might make the air take an equal share and thus halve the fall of temperature, say from a 12° fall (280–268°) to a 6° fall (280–274°) in 12 hr. Freezing-point being 273°, you see how this might prevent a frost. Still better for frost prevention are clouds. A layer of cloud at the earth's own temperature will naturally radiate at about the same rate as the earth. If unbroken, therefore, it will return all the earth's radiation and let none out at all. Cloud, however, at say 250°, duly radiating with power 20 over earth at 280° which is sending up 30, will let out 10. In short, the higher and colder or fewer are the clouds, the more heat will they let out from the earth. In any case no such preventer of frost will work if started too late. Its action is not to warm the earth up but just to retard its cooling down. Only sunshine or underground heat or artificial heat or a change of air mass will warm the earth up. In fact they are ceaselessly trying to do so. Otherwise our great cold-air masses would be even colder.

So you see how simply the earth's great air masses all depend on the single constant rate of supply of heat from the sun; and then you know how our weather types are the effects of those air masses on one another. Thus in fact you might learn your meteorology from outside the atmosphere altogether. On the planet Mars, for instance, having a nearly cloudless atmosphere with deserts, days and seasons rather like ours, yet so far from the sun that its 'solar constant' radiation figure is only of the order of half as much, its equator with midday sun overhead would receive not 30 but 15 units averaged over the day. To balance them the ground should average at least 240°, rather like our Siberian or Canadian winters. If helped by its atmosphere as much as 5 units it could average 250°, at which it would radiate 20. Not that its atmosphere would give it any more than its own rate of about 15. So it could not have more than 30 to radiate altogether. That means that it could not average a warmth greater than something of the order of 280°. Thus inferred to be between about 280° and 250°, it would not be far below water's normal freezing-point of 273° above absolute zero. Receiving at midday as much as 45 units of sunshine power together with 5 from its atmosphere whilst giving out 20 in convection, it could balance at 280° by radiation of 30. But then losing at the average rate of 20 all night it would lose 240 cal. per sq. cm. in 12 hr.; so cooling, say, 24°, it would go down to about 260° which is nearly 15° of frost. Mars, in fact, would be no more hospitable than some such place as the summit of our Mount Everest. Its actual radiation, like that of the moon, has been investigated by astronomers. Being colder away from the equatorial parts anyway, Mars is on the whole no place for hopeful interplanetary travellers.

Back on earth in lat. 50–60° the sea at midsummer receives some 30 units of sunshine power averaged over each 24 hr., together with say 10 from the air by radiation; so if just losing 5 by water evaporation it can strike a balance at 35 units with temperature about 290°. This home sea maximum temperature is roughly the same as the continental autumn- or spring-day average or midsummer minimum, but its delay is longer. In fact we might take the maritime seasons to start on those very dates (1 January, etc.) at which the continental seasons are *centred*.

4. SUMMARY

	Place no.	Date 1936	Time G.M.T.	Weather letters	Cloud				Vis. V.	Wind		Temp.	
					T.	L.	M.	H.		D.	F.	F.	C.
	01	23.9	0600	o F	8	5	–	–	0	Calm		58·3	15
	01	23.9	0700	o f	8	5	–	–	2	ESE	1	57·9	14
	18	23.9	0915	o f	8	5	–	–	2	ESE	1		
	01	23.9	1230	c f	7	5	–	–	3	ESE	2	63·0	17
	01	23.9	1300	b m	1	1	0	0	4	E	3	64·0	18
	01	23.9	1350	b	1	1	0	0	6	E	3	65·2	18
[a]	01	23.9	1600	bc	3	0	0	5	6	ENE	2	64·1	18
	01	23.9	1630	bc z_0	4	0	0	6	5	ENE	2	64·0	18
	01	23.9	2030	bc m	4	0	0	8	4	ENE	3	58·2	15
	01	24.9	0730	o m	8	0	3	–	4	SW	2	57·9	14
	18	24.9	0910	c f	7	0	3	–	3	SW	2		
	18	24.9	1140	c m	7	0	5	–	4	SW	2		
	18	24.9	1200	c m_0	6	0	5	9	5	SW	3		
	18	24.9	1600	c	7	0	7	–	6	WSW	3		
	01	24.9	1630	o	8	0	7	–	6	WSW	2	66·7	20
	01	24.9	1810	c m_0	7	0	7	–	5	SW′W	2	64·7	18
	01	24.9	2120	o $r_0 r_0$	8	0	7	–	5	SW	2	63·5	17
[b]	01	25.9	0600	o r_0	8	5	–	–	6	SW	4	62·3	17
	01	25.9	0700	o	8	7	–	–	6	SW	3	62·1	17
	18	25.9	1140	o	8	7	–	–	7	W	3		
	01	25.9	1730	c	7	0	3	–	6	W	2	64·1	18
	01	25.9	2030	o R	7	–	–	–	6	W	3	60·2	16
	01	25.9	2045	o r_0	8	–	–	–	6	W	4		
[c]	01	26.9	0645	o rr	8	7	–	–	5	NE′N	5	52·0	11
	01	26.9	1230	o	8	7	7	–	6	NE′N	2	58·0	14
	01	26.9	1440	o u	8	3	7	–	7	NE	2	62·0	17
	01	26.9	1545	c	7	0	7	–	6	NE	2		
	01	26.9	1630	c	7	3	7	–	6	NE	3	56·9	14
	01	26.9	1930	o	8	0	1	–	6	NE	2	51·9	11
	01	26.9	2140	c	6	0	7	–	6	NE	1	50·0	10

[a] Cirrus clouds moving from WNW.
[b] Cloud height only 500 ft.
[c] Rain decreasing, but going on all the morning.

SEASONS

		Tropical (30° N–30° S)	Sea or Maritime	Land or Continental	Standard Meteorological
SPRING,	starting on		1 Apr.		1 Mar.
	centred on	May	May	1 Apr.	
SUMMER	starting on		1 July		1 June
	centred on	Aug.	Aug.	1 July	
AUTUMN,	starting on		1 Oct.		1 Sept.
	centred on	Nov.	Nov.	1 Oct.	
WINTER,	starting on		1 Jan.		1 Dec.
	centred on	Feb.	Feb.	1 Jan.	

23 September is like 22 September, the equinox or end of the true sun's summer half-year, just before the end of the *continental* summer half-year with appropriate temperature maximum roughly 20° C. and day average 15° C.

Our seasons might usefully be summed up as shown on p. 73.

For the tropics, of course, the names of the seasons are better replaced by non-committal abbreviations such as M., A., N., F. (for their central months). They are quite like ours in the outer tropics, but not in the really equatorial tropics.

We might also usefully classify temperatures thus:

TEMPERATURES

	Cold °C.	Cool °C.	Warm °C.	Hot °C.
WINTER, in this country	<0	0–10	10–20	>20
SUMMER, in this country	<10	10–20	20–30	>30
TROPICAL	<20	20–30	30–40	>40

Those, you must understand, are unofficial. But you will find them remarkably easy to remember.

Here is an even fuller and more useful summary:

	Approximate temperature (generally to nearest 5°)								Moisture	
	Max.		Min.		Mean		w.p.t.		% of air	Dew-point °C.
	Abs.	C.	Abs.	C.	Abs.	C.	Abs.	C.		
TROPICS, sea air	—	—	—	—	300	27	295 to 300	25	2	24
sea air, inland	305 to 310	35	295 to 300	25	300 to 305	30	300	27	2 to 2½	25
land air, midsummer	320	50	300	30	310	40	295	20 to 25	1	10 to 20
land air, winter	300	20 to 30	280	5 to 15	290 to 20	15	280 to 20	10	½	0 to 10
HOME, land air, summer										
sea air, summer	—	—	—	—	290	15 to 17	290	15 to 20	1	15
sea air, winter	—	—	—	—	280	5 to 10	—	—	½ to 1	5 to 10
sea air, ex-tropical	—	—	—	—	—	—	285	10 to 15	—	—
sea air, ex-polar	—	—	—	—	—	—	275	0 to 5	—	—

24 September brings upper clouds from a westerly quarter which must be Atlantic sea air. Yesterday first appeared frontal high cloud types 5 and 6 which are followed to-day by increasingly frontal-looking medium cloud types 3, 5, 7.

25 September shows warm frontal type of weather with wet sea air coming right down to the ground, still raining at first with stronger south-westerly wind and much very low cloud persisting until midday and then breaking a bit for the afternoon, to be followed by heavy rain suggesting a cold front at 2030. Sure enough next day

26 September is colder, with wind right round to the north-east. So we must be upon the north side of the front.

VI

AIR MASSES

1. EQUIVALENT TEMPERATURES

	Place no.	Date 1936	Time G.M.T.	Weather letters	Cloud				Vis. V.	Wind		Temp.	
					T.	L.	M.	H.		D.	F.	F.	C.
	01	27.9	0915	o id	8	o	7	–	6	WNW	5	52·0	11
	01	27.9	1145	o ir	8	7	2	–	6	WNW	4		
	01	27.9	1230	o rr	8	o	2	–	6	WNW	2	53·0	12
	01	27.9	1400	c	7	7	7	–	6	WNW	2	53·0	12
[a]	01	27.9	1412	o m U R	8	7	–	–	4	WNW	1		
	01	27.9	1420	o m_0 t l r	8	9	2	–	5	N	3	52·7	11
	01	27.9	1500	c m/d	7	9	5	–	4	N	3	53·5	12
[b]	01	27.9	1620	b	1	3	6	o	6	N	3	50·5	10
	01	27.9	1800	b	2	o	6	o	6	N	3	47·5	09
	01	27.9	2000	b m_0	1	o	6	o	5	N	1	44·9	07
	01	27.9	2110	o m	8	–	–	–	4	N	1	47·0	08
	01	28.9	0630	b w	o	o	o	o	6	NNE	3	44·5	07
	18	28.9	1000	bc	4	2	o	o	6	NNE	4	55·0	13
	18	28.9	1200	c	6	3	6	3	7	NE	3		
	18	28.9	1345	c PR	6	3	–	–	7	NE	3		
	01	28.9	1530	bc	4	3	o	o	7	NE	5	54·9	13
	01	28.9	1600	b	2	3	o	o	7	NE	5	54·2	12
	01	28.9	1800	b m_0	o	o	o	o	5	NE	3	50·0	10
	01	28.9	2010	b	o	o	o	o	6	NE	4	46·6	08
	01	29.9	0600	bc	4	o	3	o	6	N	2		
	01	29.9	0700	c m_0	7	o	3	o	5	N	2	41·9	05
	18	29.9	0800	b m_0	o	o	o	o	5	N	1		
	18	29.9	1000	b	1	1	o	o	6	N	2	53·6	12
	18	29.9	1200	c u	7	3	–	–	6	N	3		
	01	29.9	1715	c	6	o	3	o	6	N	2	56·9	14
	01	29.9	2120	b m_0	o	o	o	o	5	Calm		47·6	09
	01	30.9	0730	o m	8	o	3	–	4	Calm		47·7	09
	18	30.9	1000	c	7	3	5	–	6	NNW	2	53·9	12
[c]	01	30.9	1430	o	8	3	5	–	6	N'E	2	57·8	14
	01	30.9	2100	o	8	–	–	–	6	N	3	53·0	12

[a] Overcast, misty, gloomy and very ugly, threatening sky, with heavy rain, bringing thunderstorm.

[b] Distant cumulonimbus remaining in the northern sky.

[c] Rather gloomy, owing to heavy cloud and London smoke.

27 September, with wind at first from a westerly quarter, shows us no longer to be upon the north side of a front in the same way as yesterday, 26 September. But at 1420 a front comes through with a bang to bring us this season's lowest temperatures yet. So it does not look like the old polar front but looks more like the arctic front.

'Polar air', we said, 'is born over ice, which makes its u.a.t. very low.' How? You know how heat may travel in three different ways:

1. CONVECTION, and/or turbulent stirring of the air by friction, the quickest process, but conditional on instability or rough flow, will even out not the actual but the *potential* temperature, potential wet temperature and potential dew-point.
2. RADIATION, slower but unconditional, evens out not potential but *actual* temperature.
3. CONDUCTION, far slower, is similar in effect, and evens out moisture (by diffusion).

In so far as arctic ice is not potentially warmer than the air, there is no convection. Radiation then tends to give all the air the ice temperature, which of course is so low that very little water vapour can be held. Arctic or polar continental (P_c) air at its source is thus very dry, cold, and tending to be isothermal.

Tropical maritime (T_m) air, on the other hand, as we also said, is warm and wet with high w.p.t., which, moreover, may have a lapse due to previous equatorial convection.

What then must happen over the western north Atlantic, for instance, in winter? T_m air moving north-east round the Atlantic subtropical 'high' must meet P_c air moving south-west round the Canadian cold winter 'high'. *These strongly contrasted air masses meet there to form the Atlantic polar front.*

But then polar air, we went on to say, 'when off the ice on to open water...naturally starts overturning with clouds feeding into it the w.p.t. from the surface'. The product, soon deep, is POLAR MARITIME (P_m) air. P_m meeting P_c or arctic air then forms what we call the arctic front.

Such are the main north Atlantic fronts. The north Pacific and other oceans have similar ones. Suitable *wind* flow patterns to make them are found in 'cols' between high-pressure regions while suitable *contrasts* naturally occur between equable *sea air* and variable cold *land air*, especially with snow lying inland to form an almost perfect heat radiator cooling down through the long winter nights. Waves and stormy cyclones developing on the main fronts themselves, however, soon complicate the pattern again, partly by shifting the high-pressure centres and partly by sweeping sea air inland, or land air to sea. Summer P_c air, though dry, is warmer, leaving less contrast, so that the fronts are often less active. T_m air, on the other hand, easily made potentially unstable, is apt to show immense power in a thundery way.

You will often have noticed the air of an English summer dawn at about 15° C. to be saturated with moisture. 1 % of the air is then water vapour. *Suppose it were all condensed.* Then it would yield enough heat to warm the air up about 25° C. The result would thus be perfectly dry air at an 'equivalent' temperature (e.t.) of 40° C. Cold air with only a third as much water vapour, saturated not at 15° C. but at freezing-point (0° C.), could in this way only be warmed through a third of 25, that is, to an e.t. of about 8° C. Tropical air, on the other hand, saturated at nearly 30° C. with 2–3 % vapour, could be warmed through 50–75°, to reach an e.t. of about 80–100° C.

Those are our extremes. The first one, with e.t. not far above 0° C. (water's freezing-point), is effectively cold, while the other, with e.t. 80–100° C. (water's normal boiling-point), is effectively hot. What is our own summer average type of air? Shall we call it effectively warm or cool? It is a border-line case. Hot desert air, you remember, has wet temperature no higher than that of our own summer dawn. As it is only upon wet temperatures that e.t. depends, such air must likewise have e.t. only around 40–50° C. That is just midway between the extreme types. May an unscientific but most descriptive term be used? We shall call it effectively TEPID.

Ordinary temperature together with moisture, determining so-called *wet temperature* which you can feel, can thus be described by EQUIVALENT TEMPERATURE effectively cold or cool or tepid or warm or hot. That is an excellent classification of types of air. If, moreover, we have first allowed for height (at standard adiabatic rates), then it will do for whole masses of air. They are effectively cold or cool or tepid or warm or hot according to their EQUIVALENT POTENTIAL TEMPERATURE. 'Effectively', of course, is an overworked word, so let us replace it by 'equivalently' or just call it 'e'. Polar winter land air, for instance, is e. cold. Ex-polar sea air is e. cool; ex-tropical sea air is e. warmer; actual tropical land air is definitely e. warm, while tropical sea air is e. hot. Polar summer air is e. cool, while tropical sea air remains e. hot, and land air between them is all e. tepid.

28–30 September shows e. very cool air of autumn polar or Arctic type, which, you notice, is *still above fog-point at only* 5° C., so that not more than $\frac{1}{2}$% water vapour is possible. Cumulus clouds appear when temperature is 12° C. With d.p.t. about 12° C. while dew-point (averaged up to the cloud base) is, say, 5° C., w.p.t. can only be about 8° C. so that e.t. is about

$$8 + \tfrac{1}{2}(25) = 20° \text{ C. approximately.}$$

As cumulonimbus clouds develop, however, we infer the surrounding air to be e. colder than 20° C. all the way up to more than 4 km.

2. SUMMARY

	Place no.	Date 1936	Time G.M.T.	Weather letters	Cloud				Vis. V.	Wind		Temp.	
					T.	L.	M.	H.		D.	F.	F.	C.
[a]	01	1.10	0700	o m$_0$	8	0	5	–	5	N	2	50·0	10
	01	1.10	2020	o	8	0	1	–	6	N	2	52·9	12
	01	2.10	0640	c m$_0$	7	0	3	–	5	N	1	50·1	10
	01	2.10	0650	bc m$_0$	5	1	3	–	5	NNE	1		
	18	2.10	1200	c	7	3	5	–	7	NE	2		
	01	2.10	1530	c	6	2	5	0	7	NE	3	55·7	13
	01	2.10	1600	bc	3	2	6	0	7	NE	3		
	01	2.10	1700	bc	2	4	6	0	6	NE	3	51·7	11
	01	2.10	1800	b m$_0$	1	5	0	0	5	NE	2	49·5	10
	01	2.10	2145	b m	0	0	0	0	4	NE	2	45·0	07
	01	3.10	0640	b	1	0	0	1	7	ESE	2		
	18	3.10	1000	b	0	0	0	0	6	SE'E	2	53·2	12
	01	3.10	1300	b	1	0	0	1	7	SE	2		
	01	3.10	1615	b	1	0	0	1	7	SE	3	55·9	13
	01	3.10	2100	b m$_0$	0	0	0	0	5	SE	1		
	01	4.10	0900	b m$_0$	1	0	0	1	5	E	1	51·5	11
	01	4.10	1030	b	1	0	0	1	6	ESE	2		
	01	4.10	1630	b	1	0	0	1	6	ESE	2	57·0	14
	01	4.10	1700	b m	1	5	0	1	4	ESE	1	55·2	13
	01	4.10	2140	b m	0	0	0	0	4	ESE	1	47·2	08

[a] C$_L$3 and C$_M$5, 7, 2 appear during the day.

1–2 October, similar to 28–30 September with normal polar-air cumulus and cumulonimbus clouds but also medium clouds suggesting an e. warmer upper air stream (possibly merely subsided under the influence of advancing high pressure), may be regarded as the beginning of the maritime autumn, the middle of continental autumn or the beginning of the winter half-year. This book, you remember, opened at the beginning of continental autumn in mid-August, and is to close at the end of maritime autumn or official midwinter. So it just covers one complete autumn, of which we are now in the middle. All the main ideas introduced thus far, through thermodynamics, potential wet and equivalent temperatures, from types of weather to types of air, may now be summed up concisely before we turn to hydrodynamics, the theory of wind and synoptic charts.

Polar winter land air is e. cold: ex-polar sea air is e. cool; ex-tropical sea air is e. warmer, tropical land air is e. warm, and equatorial or tropical sea air is e. hot. Polar summer air is e. cool; tropical sea air remains e. hot, and land air between them is just e. tepid.

Now in mid-autumn, midway between the summer and winter average temperatures, we should therefore expect to find:

1.	ARCTIC	(A)	air e. cold;
2.	POLAR CONTINENTAL	(P_c)	air e. very cool, rather cold;
3.	POLAR MARITIME	(P_m)	air e. cool;
4.	TROPICAL MARITIME	(T_m)	air e. warm;
5.	TROPICAL CONTINENTAL	(T_c)	air e. very warm, rather hot;
6.	EQUATORIAL	(E)	air e. hot.

Then if the air all stayed still, the boundaries of these air mass types would naturally be found:

between 1 and 3, around polar seas as the ARCTIC FRONT;
between 2 and 3, around polar lands as the ARCTIC FRONT
(modified);
between 3 and 4 as the POLAR FRONT;
between 4 and 6, around tropical seas as the EQUATORIAL FRONT;
between 5 and 6, around tropical lands as the EQUATORIAL FRONT
(modified).

In so far as the air does not stay still, these fronts shift about, and so the air masses get warmer or cooler in moving to warmer or cooler places. So we get characteristic weather types, generally:

FINE in continental (C) air;
FAIR OR SHOWERY in maritime (M) air getting e. warmer, such as P_m air anywhere outside the tropics, or T_m air anywhere inside the tropics;
CLOUDY OR FOGGY in maritime air getting e. colder, such as T_m air anywhere outside the tropics, or anywhere over high ground;
RAINY in maritime air anyhow disturbed, such as by ascent on fronts or mountains;
STORMY in maritime air violently heated or disturbed, e.g. by a frontal wave or cyclone.

VII

AIR HYDRODYNAMICS

GRADIENT WIND AND FRONTS

Air thermodynamics, you will remember, was introduced by quotation from Chapter I.

'Our analysis of...the way in which properties change as the air goes along is largely a matter of thermodynamics....Analysis, on the other hand, of...the way in which properties are bodily carried along by the air-flow is largely a matter of *hydro*dynamics, the theory of fluid flow. That is part of mechanics. It speaks of density, pressure and circulation or flow... circulation or flow is a matter of wind....Wind is already one of the four weather elements we have observed.' So let us go on to quote our commentary from the end of 20 August.

'There is indeed more in the beautiful patterns of air-flow than meets the eye...and we turn to the subject of wind.

'How does water flow over land? It depends on the lie of the land, that is, the distribution of *height* of the land, which is mapped out by *contours*.

'Likewise at any one height the distribution of *pressure*, mapped out by *isobars*, determines the air-flow or wind...a matter of pattern.

'Suppose there is for a while no change in the pattern. Then apart from friction we recognise:

'**Pressure gradient force** from high to low pressure...measured by the pressure gradient or isobar spacing, just as the spacing of contours measures the gradient or slope of the land.

Geostrophic force to the right of the wind direction in northern latitudes, or to the left in southern...and

Centrifugal or centripetal force whenever the wind—or anything else—moves in a curve instead of in a straight line.

'If the wind flows straight, we can disregard centrifugal or centripetal force, and therefore say that when all is steady the first two forces alone must balance each other. As the geostrophic force depends on the wind, while the pressure-gradient force depends on the pressure gradient, the steady wind that blows when they balance can only depend on the pressure gradient which the isobars show. That is the GEOSTROPHIC WIND.'

'Motion', to quote first principles once again, 'is purely relative. We think of the ground as being at rest while the atmosphere does the turning. In northern latitudes as the earth turns anti-clockwise relative to the heavens, so must the air or heavens turn clockwise...relatively to the earth.'

The earth turns on its axis at 1 revolution per day. If you were at the north pole, for instance, in latitude 90° so that the axis was in the same direction as your local vertical (at right angles to horizontal), then the ground (or ice) beneath your feet would be turning *horizontally* round your vertical line at 1 revolution per day. But if you were at the equator, in latitude 0°, so that *to you* the earth's axis appeared no longer vertical but horizontal (north-south), then the ground beneath your feet could not be turning horizontally round your vertical line at all. The sun's, moon's and stars' daily tracks across the heavens, for instance, or rather their shadows or projections upon your horizontal ground- or sea-level, would all run straight east-west without horizontal curvature.

In latitude ϕ (ϕ being our usual symbol for latitude), in short, the ground only turns *horizontally* under the heavens at a rate of sin ϕ revolutions per day. That is zero at the equator where ϕ is 0, but is 1 at the poles where ϕ is 90°. Readers, of course, are assumed (or asked) to be familiar with sines as well as with a few other mathematical terms we shall have to use.

Twice this rate, that is 2 sin ϕ revolutions per day, is given a special symbol too. To meteorologists it is the Greek letter λ. As a special concession to general readers, however (the only one of its kind now called for at all in this book where so little mathematical language is used), the corresponding letter of our more familiar alphabet will be used. That is L.

So-called calm air, for instance, is really circulating horizontally with the ground at a rate $\frac{1}{2}L$. We must remember, of course, to say what we mean here by 'really'. We mean *relatively to the outside heavens*. We can then say that air which is really going straight must appear to us on the ground to be moving in part of a circle (see Fig. 16). In one unit of time after starting eastwards from A, say, the air has arrived over B. Relatively to the ground, you see, its 'real' straight track (AB) is not now due east but is turned through angle $\frac{1}{2}L$ to the south of east. The direction of the air's motion itself, also relatively to the ground, thereby appears turned through twice this angle, namely L. Whilst unaltered in *speed*, say V, the air's velocity has been changed in *direction* through L radians. From Fig. 17 you then see that the change of velocity is L times V. But CHANGE

OF VELOCITY IN UNIT TIME means ACCELERATION, and acceleration is FORCE ON UNIT MASS. So on unit MASS of air there is GEOSTROPHIC FORCE LV. Finally, therefore, on unit VOLUME of air, whose mass is simply the air's DENSITY which we call ρ (Greek 'r', not unlike our 'p', but so widely used in science that we here draw the line at making any further concession to general readers), the geostrophic force is ρLV.

Now the air pressure is denoted by our own letter 'p'. The pressure gradient then is called 'grad p', denoted by ∇p. That is the PRESSURE CHANGE PER CENTIMETRE or unit length. But pressure itself is force per square centimetre or unit area. So ∇p is force on one *cubic* centimetre or

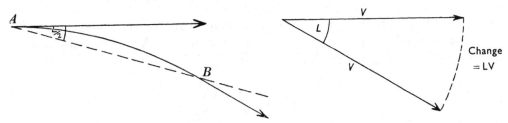

Fig. 16. Air moving over rotating earth.

Fig. 17. Acceleration relative to rotating earth.

unit *volume*. We have just seen that on unit volume the geostrophic force is ρLV. The geostrophic wind is the wind speed (V_{gs}) deduced by assuming the balance of these two forces alone, when $\rho LV_{gs} = \nabla p$. Thus

$$V_{gs} = \frac{\nabla p}{\rho L}.$$

That is our simple formula for GEOSTROPHIC WIND SPEED.

The third possible force, you remember, is the centrifugal or centripetal force recognised whenever the air—or anything else—moves in a curve instead of in a straight line. Again, of course, it is purely relative to our standard of motion, whatever it is. But we must be consistent. As long as we reckon all relatively to the earth we must duly reckon the geostrophic and centrifugal forces equally real whenever the wind blows over the earth. Outsiders in space might call them fictitious, but our own feet are on earth just now. When anything moves with speed V round a curve with radius r instead of in a straight line, you will know (or can anyway see for yourself if you draw it on paper and recollect your school geometry) that the change of direction of motion in a unit of time will change the resultant *velocity* by the amount $\dfrac{V^2}{r}$. That is the *centripetal* acceleration, so called because it acts towards the centre of the curve, i.e. '*seeks*' *the centre.*

So on unit volume of air there is a centripetal force

$$\rho \frac{V^2}{r}.$$

The geostrophic force $\rho L V$ and the pressure-gradient force ∇p, which is $\rho L V_{gs}$, will no longer balance each other but have a difference $\rho L\,(V - V_{gs})$, which may be called the *a*geostrophic force, equivalent here to the centripetal force $\rho \dfrac{V^2}{r}$ so that $L\,(V - V_{gs}) = \dfrac{V^2}{r}$, which is an equation telling us how strong (V) the wind must now be when all three forces are balanced. That is defined as the GRADIENT WIND, denoted by V_{gr}. The equation's only solution that is applicable in practice is

$$V_{gr} = \tfrac{1}{2} r L \left(1 - \sqrt{1 - \frac{4 V_{gs}}{rL}} \right),$$

reckoning r to be negative for cyclonic but positive for anticyclonic curvature of air-flow. 'Cyclonic', of course, means anti-clockwise in northern latitudes but clockwise in southern.

That is a formula for a fictitious but often almost true wind simply in terms of

curvature of flow (represented by radius r),
latitude (represented by factor L),
and pressure gradient (represented by geostrophic wind speed V_{gs}).

V_{gs}, for which you remember the simpler formula

$$V_{gs} = \frac{\nabla p}{\rho L},$$

is only a first approximation to the true wind. It differs from V_{gr} by

$$V_{gr} - V_{gs} = \frac{V^2_{gr}}{rL},$$

which is known as the CYCLOSTROPHIC component of the wind. With r it is reckoned negative (backwards against V_{gs}) if cyclonic but positive if anti-cyclonic, being zero, of course, if the flow is perfectly straight.

84

Now we have to describe the idea of a FRONT in hydrodynamical language too. How are we to do it?

What exactly does gradient mean?

Gradient of LAND simply means rate of variation of height (say h) with horizontal distance along the direction of steepest slope, that is at right angles to the contours. So we may call it *grad h*, we may write it ∇h, and we may measure it by the spacing of contours marking out unit or standard intervals of h.

Likewise ISOBARS (at unit intervals of pressure p) give ∇p; ISOTHERMS (of temperature T) give ∇T; and ISOSTERES (of so-called SPECIFIC VOLUME $\alpha = 1/\rho$, the volume per unit mass, of air or of anything else) give $\nabla \alpha$.

Isosteres, you see, are contours marking out *density*, but at unit intervals not of density itself (say in gm. per c.c. or lb. per cu. ft.) but of its reciprocal (c.c. per gm. or cu. ft. per lb.). In so far as they are not the same as isobars of pressure, they form a network or *grid* with the isobars. The meshes of this network are called SOLENOIDS.

These are useful conceptions. The steeper the gradients $\nabla \alpha$ or ∇p, i.e. the more closely crowded are the isosteres or isobars, and the more they differ in direction and thus intersect to form many meshes, the more meshes will there be in any unit of area. The number of meshes per unit of area can then be called the *solenoid concentration* and most concisely expressed mathematically as the *vector* product

$$\nabla p \times \nabla \alpha.$$

Now the CIRCULATION round any circuit is the velocity integrated with respect to distance along the circuit. The circulation then per unit area of closed circuit is what is known as VORTICITY, twice the angular velocity or turning rate. A celebrated hydrodynamical theorem tells us that the *rate of increase of circulation round any closed circuit that moves with the fluid is proportional to the solenoid concentration*.

Suppose there are *no solenoids*. Then the rate of increase of circulation is zero. In other words the circulation is constant, say C. To a circuit of area A the earth itself, with turning rate $\frac{1}{2}L$ and thereby vorticity L, must obviously contribute A times L to the air's total circulation C. The rest of the air's vorticity, say R, must be *relative to the earth*. Such is the actual wind or air-flow with which we are concerned.

So with $A(L+R) = C$ which is constant, any increase of L must decrease A or R. In other words any air going to higher latitudes must contract horizontally (expanding upwards if not downwards) or else spin less (cyclonic way round). Or it may do both. The less it decreases its cyclonic spin, the more it must contract horizontally. Or if decreasing in L, it must increase in A or R. In this case it can only contract horizontally if increasing too much in cyclonic spin.

A straight air current, in particular, will expand or *diverge* horizontally when moving equatorwards, but contract or *converge* when moving polewards. In the first case, therefore, it tends to subside with clear weather, whereas in the second it tends to be cloudy. Being geostrophic too, if still without solenoids, so that ρ is constant down-wind along the isobars, it has no other reason to converge or diverge; for, having speed (V_{gs}) in inverse proportion to the isobar spacing $\left(\dfrac{1}{\nabla p}\right)$, it flows between any two isobars just as water flows through a pipe, unable to pile up anywhere.

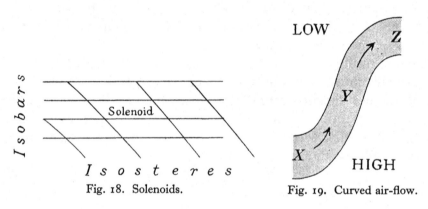

Fig. 18. Solenoids.   Fig. 19. Curved air-flow.

A curved air current, on the other hand, as in Fig. 19, is not just geostrophic but has a *cyclostrophic* component, which in northern latitudes would act backwards at X, where the curvature is cyclonic, and be zero at Y and then act forwards at Z. From the ends of this 'pipe' it would thus flow out, so that of the total gradient wind you see less in at X than out at Z. That is a case of horizontal divergence, as might equally well be deduced from the big decrease of cyclonic spin (R) with only a small increase of L. $L+R$ altogether decreases whilst $A\,(L+R)$ is constant (C), so the air's horizontal area A increases. That simply means horizontal divergence.

If the isobars themselves diverge down-wind on the right hand or starboard side of an air stream whilst converging down-wind on the port side as in Fig. 20, the ageostrophic wind components again act outwards to give the total wind horizontal divergence. Combining Figs. 19 and 20 in Fig. 21, a kind of pattern often seen in practice, we infer horizontal divergence from the space in between the pressure 'trough' and the 'ridge', with corresponding convergence into the space from the ridge to any trough further down-wind, so that the air as a whole is piled up down-wind of the ridge. This raises its pressure and so moves the ridge down-wind.

That is the kind of way in which 'highs' and 'lows' are moved. Round a stationary cyclone the track of the wind may be equally curved on all sides, not only relatively to the cyclone centre but relatively to the earth underneath. But if the whole cyclone is moving at all so that we can define its starboard side, then obviously in northern latitudes, as its starboard winds must blow the same way as the cyclone centre is going, they curve less over the earth underneath than the opposite winds on the port side.

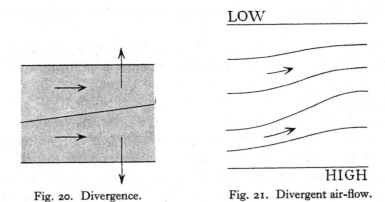

Fig. 20. Divergence.   Fig. 21. Divergent air-flow.

So their cyclostrophic components (acting backwards, *against* the cyclone's drift) are smaller than the port-side cyclostrophic components which duly act *with* this drift. Air is accordingly piled up ahead of the cyclone, raising the pressure, or rather making it fall less fast, and thereby slowing down the low-pressure centre's advance; for here you see the air moving over from starboard to port side, thus increasing its cyclonic spin (R) and so decreasing its area (A) by horizontal convergence.

In southern latitudes several of these factors apply the other way round, but as an even number of negatives may be combined as a positive factor we have the same final effect of gradient winds converging ahead whilst diverging behind a moving cyclone. Likewise, of course, they diverge ahead whilst converging behind an anticyclone. In either case they naturally help to slow down the drift of the centre, though if they are not the true winds they may not succeed.

87

'Head-on FRICTION', you will remember, 'slows the wind down and deflects it across the isobars towards lower pressure', while 'RISING OR FALLING PRESSURE, which changes the pattern, has an effect of pushing the air across the isobars towards the region of most rapidly falling (or least rapidly rising) pressure.' Indeed the very effect we have just discussed, of gradient wind itself trying to raise the pressure by convergence or lower it by divergence, will duly create these extra (non-gradient) wind components. We must take care, however, to distinguish what is relative to the earth from what is only relative to the moving isobars.

'Combined with the drift across the isobars due to friction it causes out-flow or divergence from centres of high or rising pressure, with convergence into low or falling pressure. To make up for excessive horizontal convergence the air must rise, whereas to make up for excessive divergence it sinks; and you know how much the air's rising or sinking affects the weather.'

Thus far we have neglected our solenoids. We supposed that there were none. That, however, can only be true if pressure is uniform (implying no wind), or if density is uniform (implying no fronts), or if pressure-lines run with density-lines.

Solenoids, in short, being the isobar-isostere network's meshes, can only vanish if there are either no isobars, or no isosteres, or no intersections. Absence of intersections simply means constant air density along any isobar. That implies *constant density of the gradient wind*. If the air is warmed or cooled as it goes along, then there must be solenoids after all.

Remember that air is a fluid. Fluid motion is easy enough to feel, or to see approximately, but not so easy to see exactly, or at any rate not so easy to *follow* as solid things are. You recognise anything solid as being com-posed of the same material particles all the time. Not just the chief properties but the *identity* of its material is preserved. That means no mixing with others. It means letting none in or out through its boundary surface. Think of a mass of air just like that. Never mind the transparent fluid nature that makes it so elusive to mark out exactly. It is no less definite than, say, a mass of oil in water. That, too, is all fluid. But it is easy to watch if coloured.

Our air is defined, then, not to mix with other air, nor to let any in or out through its boundary surface; or rather its boundary surface may be *defined* as that through which nothing flows in or out.

If it shrinks horizontally, then it must deepen vertically. How much? To keep total volume constant? Not if it is heated at the same time, for that would itself enlarge the total volume. What, then, must remain constant? Surely the total *mass*. The volume will only do so as well if the density also stays constant.

In this particular case we can simply say that the total horizontal *inflow* of fluid into any particular space must be exactly balanced by the total *outflow* up and down. Outflow can always be counted as inflow but with a 'minus' sign. If we allow for heating (inconstant density) then we must speak of the flow not just of the fluid *material* but of its mass, or rather of its *momentum*.

Over level ground or sea, of course, the bottom of this mass of air is bound to have average flow horizontal, neither upward nor downward; so any vertical outflow must be upward, aloft. But that is what causes bad weather. So to account for most of the weather we have only to study the air's HORIZONTAL CONVERGENCE. Already we know (for constant density) how it increases with latitude and cyclonic spin. Those were our L and R. For inconstant density it is the horizontal convergence not of the wind itself but of its momentum that has to account for the weather. Air being heated, for instance, must expand upward if not horizontally. Even with momentum still perfectly balanced horizontally you see that the wind itself then cannot be so. The heating may be of the air mass itself as it goes along, or else of some fixed place as warmer air comes from elsewhere. In either case some air has to rise. L, R and these two different forms of heating make altogether four simple hydrodynamical factors affecting the weather. We have worked out the effect of $L+R$ without solenoids; but Nature adds the other two factors with solenoids, one of them being in the already familiar form of FRONTS.

'A front', we said, 'is simply the boundary between any two different masses of air', or rather two lots of air which differ in *density*, 'whether sudden or only gradual in transition. With the effectively warmer air always sloping up over the colder air mass, the front is bound to share their motion. If the air-flow is parallel to the front, then obviously the front stays still, quasi-stationary.

'But if the warm air overtakes the colder air at an angle, even without any "total" change taking place in either mass of air as it goes along, then the bodily transport or advection of the new air is bound to appear as a change of air at any one place underneath. Whether sudden or only gradual, this is the *local change* caused by the movement of a front.'

Total change may in fact be included (under the heading of 'gradual transition') as a diffuse quasi-stationary front to be combined with the main moving front. The result is a slightly different pattern of front which is rather diffuse but nevertheless covers both our solenoid factors:

(i) heating of each air mass itself as it goes along, and

(ii) heating of any fixed place underneath it as a warmer air mass comes from elsewhere.

The weather depends, then, upon how sudden or gradual the frontal transition is. 'Born as two air masses meet, a front will intensify if the air-flow pattern is such as to squeeze the air masses together and make the transition zone narrow so that all "contour" lines which mark out the different air-mass properties are closely crowded.' Isosteres, for instance. *Isosteres* crowded across isobars, as you already know, form *solenoids*. 'A change of flow pattern which draws them apart again will weaken the front's activity.' Naturally, it reduces the solenoid concentration and thus the rate of increase of circulation, the rate of change of wind at the front. And so our quotation from an earlier chapter has brought us back to the motto of our air's hydrodynamics:

'There is indeed more in the beautiful patterns of air-flow than meets the eye.'

As a front is defined to be the boundary—strictly speaking, the boundary *surface*—between two air masses, neither air mass can flow through it. Any velocity difference between the front and the air itself must therefore be parallel to this surface. At sea-level in the cold air under the front no such flow can be downwards, or it would enter the sea. If not upwards either, then it must be horizontal. But this air is naturally in the form of a wedge, whose top is the frontal surface, whose end is the front itself at sea level. Obviously it has no room at all to pile up horizontally here, while if it tried to retreat from the front it would leave an impossible vacuum. So it must move with the front. To put it the other way round, *a front (not too high above sea-level) must move horizontally with the colder air stream.*

Any difference then between the cold and warm air velocities is simply the velocity difference between the *warm* air stream and the front itself. Being parallel to the boundary surface, which is not quite horizontal, it must generally have an upward or downward component. In this we are interested. It is computed, of course, by multiplying the slope of the surface (say s, which is of the order of $1/100$) by the horizontal velocity difference, in the same direction. Even purely geostrophic winds according to the formula

$$V_{gs} = \frac{\nabla p}{\rho L}$$

will differ at least in density ρ on opposite sides of a front. This shows *momentum* still perfectly balanced horizontally, yet *velocity* not so.

At unit distance ahead of the front at sea-level the cold-air wedge has depth s. With average density ρ it therefore exerts a pressure ρs, while the same depth of warm air (mean density ρ') over the front at unit distance away is exerting a pressure $\rho's$. To the *pressure gradient* (pressure difference over unit horizontal distance), the front itself contributes $(\rho-\rho')$ times s. So we infer a geostrophic wind velocity difference of about

$$\frac{(\rho-\rho')\,s}{\rho L}$$

between the cold air and the warm. That is the frontal wind SHEAR.

Observant readers will object that something is missing, or rather implied but not explicitly shown in the formula. That is the gravity factor called g. It will be instructive to digress here on how it arises and how we detect its omission.

ρ stands for density, say in gm. per c.c. The density difference $\rho-\rho'$ is of course also in gm. per c.c. So $\frac{\rho-\rho'}{\rho}$ is in gm. per c.c. divided by gm. per c.c. which duly cancel out to leave neither mass nor volume nor length nor time nor any other dimensional unit but just a pure number.

s, too, is just a pure number, being *height* (say in cm.) divided by *horizontal distance* also in cm. The length unit (cm.) again cancels out. Our formula's only remaining factor is L. What is that? It is a fraction of the earth's rate of turning, say in revolutions per day. Now a revolution means width (or radius) divided into distance swept round in a circle. Both may again be reckoned in cm., which thus cancel out to leave a pure number. Nothing is left but pure numbers excepting the words 'per day'.

A day is a unit of time, so that $\frac{(\rho-\rho')\,s}{\rho L}$ is not a pure number but a *time*, which is certainly not the same thing as velocity, whether of wind or anything else. Yet the formula is supposed to give a wind velocity difference, say in cm. per day, or more conveniently in cm. per sec. 50 cm. per sec., for instance, is about 1 KNOT, the familiar unit of wind speed.

By what, then, must we multiply *time* (say in seconds) in order to get *velocity*, say in cm. per sec.? Obviously by something in cm. per sec. per sec., briefly written as cm. per sec.[2]. What kind of quantity is expressed in these particular units?

Seeking what is lost, let us go right back to the formula

$$V_{gs} = \frac{\nabla p}{\rho L}$$

to see whether even this is correct. What does ∇p stand for?

Pressure gradient. What is pressure gradient? It is pressure change per cm. or unit length.

'But pressure itself', we said, 'is force per sq. cm. or unit area. So ∇p is force on 1 c.c. or unit volume. We have just seen that on unit volume the geostrophic force is $\rho L V$,' and those two forces per unit volume are supposed to be equal.

ρ, being density, can be expressed in gm. per cm.^{3}

$1/L$, being time, can be expressed in sec.

V, being velocity, can be expressed in cm. per sec.

Then $\rho L V$ is in $\dfrac{\text{gm. cm.}}{\text{cm.}^3 \text{ sec. sec.}} = \dfrac{\text{gm. cm.}}{\text{sec.}^2}$ per unit volume (cm.^{3}).

But it is *force* per unit volume. Therefore force must be in $\dfrac{\text{gm. cm.}}{\text{sec.}^2}$.

Newton's law states that force is mass times acceleration. Acceleration is *velocity* at a time rate, say, *per sec.*

Velocity is distance per time, say *cm. per sec.*

Then MASS×ACCELERATION is in GM. CM. PER SEC. PER SEC.$= \dfrac{\text{gm. cm.}}{\text{sec.}^2}$.

Those are units of force.

The *weight* of a gm. mass (say 1 c.c. of water) under our normal gravity is then 1 gm. multiplied by the normal *gravity acceleration*

$$g = 981 \frac{\text{cm.}}{\text{sec.}^2}.$$

That is our lost factor.

'At unit distance ahead of the front at sea-level,' we said, 'the cold-air wedge has depth s,' measured in cm., say. With average density ρ $\left(\text{say in } \frac{\text{gm.}}{\text{cm.}^3}\right)$ it therefore has mass ρ gm. over each sq. cm., whose *weight* is ρ times the weight of a gm. mass. That is $\rho \times g$ $\frac{\text{gm. cm.}}{\text{sec.}^2}$ over each sq. cm. So its pressure, in basic units $\left(\frac{\text{gm. cm.}}{\text{cm.}^2 \text{ sec.}^2}\right)$, is not ρ but $g\rho$. Likewise we replace

ρ' by $g\rho'$, so that the wind shear is

$$\frac{g\,(\rho-\rho')\,s}{\rho L},$$

which now correctly has velocity units $\left(\text{say}\ \dfrac{\text{cm.}}{\text{sec.}}\right)$.

To beginners this will have shown the useful idea of DIMENSIONS, by which we can always test, for example, whether a formula is incomplete or upside down. Length, mass and time, abbreviated to **L**, **M** and **T**, have hitherto been accepted as the fundamental so-called dimensions from which everything else in physics may be derived. Velocity has the dimensions of $\dfrac{\mathbf{L}}{\mathbf{T}}$; acceleration $\dfrac{\mathbf{L}}{\mathbf{T^2}}$; force $\dfrac{\mathbf{ML}}{\mathbf{T^2}}$; pressure $\dfrac{\mathbf{ML}}{\mathbf{L^2T^2}}=\dfrac{\mathbf{M}}{\mathbf{LT^2}}$; energy=force $\times$ distance$=\dfrac{\mathbf{ML^2}}{\mathbf{T^2}}$; power=energy $\div$ time$=\dfrac{\mathbf{ML^2}}{\mathbf{T^3}}$.

In particular, therefore, the square of the square of the absolute temperature of a perfect radiator, being proportional to the radiation power as we already know, has dimensions $\dfrac{\mathbf{ML^2}}{\mathbf{T^3}}$, so that the temperature itself might be said to have the dimensions

$$\frac{\mathbf{M^{\frac{1}{4}}L^{\frac{1}{2}}}}{\mathbf{T^{\frac{3}{4}}}}.$$

It all depends on what you mean by temperature. As the temperature already defined otherwise has been found so related to radiation power, we might define it again by this power instead. Either way, you see, we are using observed laws or regularities found in Nature. So did Newton when he defined force.

But now it is time to return to the subject of fronts and their wind shear. This measures how fast the warm and cold air streams rub across each other (parallel to the front) whilst advancing at right angles to it. 'In a frontal wave depression or cyclone,' we said in our earlier chapter, 'you know that the different air masses have different flow. The different flow of the different air masses implies a different run of the isobars, which therefore kink at the fronts.' The kink is just the combined effect of a continuous upper-air pressure gradient with our extra one $g\,(\rho-\rho')\,s$ which only exists in the cold-air wedge. The resultant gradient is thus discontinuous at sea-level, even though the pressure itself is not so. No isobar can possibly break, either at a front or anywhere else; so it crosses the front at only one point at a time. Neighbouring isobars cross the front at points whose spacing, in fact, gives the pressure-gradient component along the front and thus gives the geostrophic wind component at right angles, which we may call the front's geostrophic rate of advance, say V_{fgs}. At right angles to this, however, we find our wind shear

$$\frac{g\,(\rho-\rho')\,s}{\rho L}.$$

You see how it varies with

$\rho-\rho'$, the sharpness of air-mass contrast;

s, the steepness of the frontal surface, which in fact may be limited by the tremendous stirring and *mixing* effect of any excessively strong shear of this kind;

and L, which in low latitudes likewise makes the shear so strong for our familiar values of s that only less steep fronts will survive.

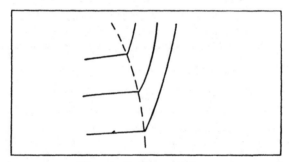

Fig. 22. Isobars at a front.

'Ahead of a warm front the warm air is, of course, deepening over any fixed place, whereas behind the front it can deepen no further. So there must be an abrupt change of rate of fall of total pressure, as is implied by the kink in the isobars passing by. Behind a cold front, too, cold air comes in an ever-deepening wedge, with the same effect.' At right angles to the front, the pressure-gradient components differ by $g\,(\rho-\rho')\,s$; accordingly if the front's actual rate of advance is V_f, so that a unit of distance is covered in time $1/V_f$, the front itself must be contributing $g\,(\rho-\rho')\,sV_f$ to the rate of change of sea-level pressure.

The *net* rate of change of sea-level pressure is called the BAROMETRIC TENDENCY, usually measured in mb. per 3 hr. In basic units it would be

$$\frac{\text{gm.}}{\text{cm.}^2\,\text{sec.}^2}\frac{\text{cm.}}{\text{sec.}}=\frac{\text{gm.}}{\text{cm. sec.}^3},$$

whose dimensions are $\dfrac{\mathbf{M}}{\mathbf{LT^3}}$. g is $\dfrac{\mathbf{L}}{\mathbf{T^2}}$; $\rho-\rho'$ is $\dfrac{\mathbf{M}}{\mathbf{L^3}}$; s is pure number, while V_f is velocity $\dfrac{\mathbf{L}}{\mathbf{T}}$; so $g\,(\rho-\rho')\,sV_f$ has the same dimensions $\dfrac{\mathbf{M}}{\mathbf{LT^3}}$ as barometric tendency.

Now $g\,(\rho-\rho')\,s$, being the difference of pressure-gradient components across the front, is simply measured by the difference of average isobar spacings along any line drawn at right angles to the front, say between 0·01 mb. per km. on one side, and −0·02 on the other. Two minuses make a plus, so the true 'difference' is 0·01+0·02=0·03 mb. per km. That is $g\,(\rho-\rho')\,s$.

Meanwhile we can just as easily compare the average tendencies on the two sides, say −1 mb. per 3 hr. and +2, whose difference is 3 mb. per 3 hr. =1 mb. per hr. That must be $g\,(\rho-\rho')\,sV_f$.

So we deduce the true speed of advance of the front or trough-line to be in this case

$$1\,\frac{\text{mb.}}{\text{hr.}}\div 0\cdot 03\,\frac{\text{mb.}}{\text{km.}}=33\ \text{km./hr.}$$

to the nearest unit.

That is the useful 'tendency' method of estimating its speed. You can use it not only for fronts and troughs but for 'lows' and 'highs', whether frontal or not. You can even judge their acceleration; for, being defined as rate of change of velocity, it will naturally be shown by any *steepening* of the barometric *tendency* gradient or by any *blunting* of the isobars' sharp kinks or smoother curves. Measurement of V_{fgs}, on the other hand, is another method, the 'geostrophic' method. Yet a third method is extrapolation from history, *extending the path of previous actual plots of a front or trough-line or pressure centre.* Being graphical, it is often by far the best method. Obviously it allows for acceleration, retardation or other developments. At a glance, in short, this 'path' method shows not only all that the other two methods describe in figures, but far more than any quick calculation could solve at all.

With these limitations, how do the methods compare in effect? You will see for yourself as we go through synoptic charts in practice. All these hydrodynamical ideas just introduced will now be developed with illustrations, culminating in a non-mathematical outline of the isentropic and isobaric upper-wind forecasting methods developed in recent years, together with the writer's technique and simplified gradient-wind scale.

VIII

SYNOPTIC CHARTS

1. INTRODUCTION

	Place no.	Date 1936	Time G.M.T.	Weather letters	Cloud				Vis. V.	Wind		Temp.	
					T.	L.	M.	H.		D.	F.	F.	C.
	01	5.10	0700	b f w	1	0	0	1	3	Calm			
	01	5.10	0730	b f w	1	0	0	1	3	E'S	1	43·1	06
	01	5.10	0800	o f w	8	5	–	–	3	E'S	2	45·0	07
	01	5.10	0830	o f $d_0 d_0$	8	7	–	–	3	ESE	3		
	18	5.10	0850	o m $r_0 r_0$	8	7	–	–	4	ESE	4		
	18	5.10	1050	c m_0	7	0	5	–	5	E	4		
	18	5.10	1145	c	6	2	–	–	6	E'N	5		
	18	5.10	1400	c	7	4	–	–	6	E'N	5		
	18	5.10	1500	c	7	4	–	–	6	E'N	5		
	01	5.10	2120	bc	4	0	3	0	6	E'N	3	43·9	07
[a]	01	6.10	0730	bc m_0	3	0	3	0	5	NE	3	41·5	05
	18	6.10	1000	bc	6	1	0	4	6	NE'E	3		
	18	6.10	1330	c r_0	7	3	–	–	6	ENE	3		
	01	6.10	1830	b m_0	1	0	3	0	5	ENE	3	45·0	07
	01	6.10	2220	b m_0	0	0	0	0	5	ENE	3	42·0	06
	01	7.10	0720	bc	5	3	3	0	6	N	2	43·6	06
	01	7.10	0740	b m_0	1	0	5	0	5	N	2	44·1	07
	18	7.10	1100	c z_0	7	0	7	–	5	NE	4	50·2	10
[b]	01	7.10	1540	c	7	2	6	–	6	NE'N	2	49·5	10
	01	7.10	2140	b m	2	0	3	0	4	NNE	1	45·0	07

[a] Mist 500 ft. deep.
[b] Large cumulus nearly becoming Cb but then spreading out without any showers.

Here we have synoptic charts. These are small reproductions copied by hand from Air Ministry Meteorological Office *Daily Weather Reports* (*D.W.R.*), International Section.

There are four standard times daily:

Midday 13 h. G.M.T.: inset, top left-hand corner.

Evening 18 h. G.M.T.: main chart on the left.

Midnight 01 h. G.M.T.: inset.

Morning 07 h. G.M.T.: main chart on the right.

Those are the Synoptic Hours. Nowadays the standard hours are 12, 18, 00 and 06 G.M.T. instead. Meteorological Office working charts are

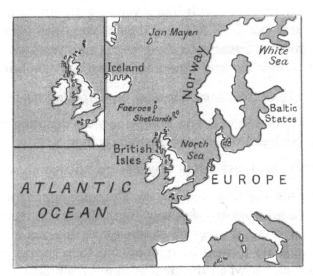

Fig. 23. Key map.

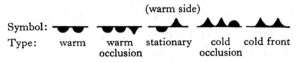

Type: warm   warm   stationary   cold   cold front
            occlusion   occlusion

Fig. 24. Symbols for fronts.

also done at the intermediate third hours, making altogether eight daily. Nearly all countries agree to report at these times.

The official picture is shown, as drawn up at the Air Ministry Central Forecasting Office. Background is an outline key map, on which observations are plotted. The number of weather elements shown depends somewhat on the scale of the map. On our small charts are shown arrows flying with the WIND (not *into* it like a vane), with circles denoting calm. On bigger charts (as *D.W.R.*) are shown WEATHER (symbols); WIND (arrows complete with feathers denoting force), and TEMPERATURE (°F.). On Meteorological Office working charts are shown full synoptic data.

ISOBARS are the thick, full lines, showing pressure and air-flow.

FRONTS are the lines adorned with bumps or teeth denoting type (whether warm or cold, etc.) and which way they are going.

AIR-MASS TYPES or WEATHER TYPES may also be labelled on working charts.

5 October ends the fine weather by clearance of an orthodox night radiation fog in a most unorthodox way. With slight rain reported between 0830 and 0900 it is swept away by fresh easterly breezes in which plenty of cumulus clouds are formed by midday and then spread out. Contrast with only 24 hr. earlier when the air was so mild, almost cloudless, totally different.

Our very first charts show the front going through from the dying whirl of an old occluded cyclone over the Baltic States sweeping the winds round from Russia, Lapland and northern seas to displace the mild south-easterlies over Britain and France. They show the high-pressure ridge too, and how it spreads down with the rising pressure behind the front. They also show the 'high' surrounded by 'lows'—the Baltic one dying, a Mediterranean one being born, an Arctic one complete with warm and cold fronts just appearing at the top of the picture on its way eastward (round the 'high'), and sundry Atlantic ones which show no development.

Our own new air mass, with clouds spreading out in quite a familiar way, has obviously an e. warmer air stream above it. Is it merely subsided air from the high-pressure ridge? Perhaps so, for the ridge is certainly coming, and the upper air is not full of cloud like really wet maritime air. Nevertheless the medium clouds' persistence might be significant, especially since they had first appeared before the cumulus so that they cannot have been formed solely by cumulus clouds spreading out. The front, having cold-front form and thus sloping up backwards, forms a kind of damp

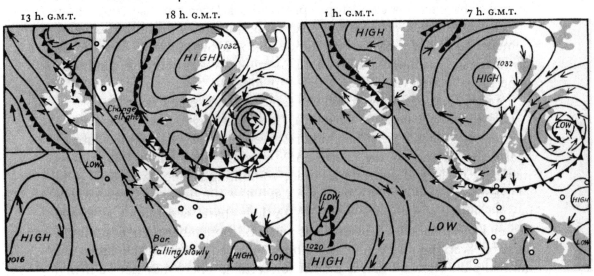

Fig. 25. Cyclone or depression sweeping north-easterlies from the Baltic across the North Sea to Britain after south-easterlies.

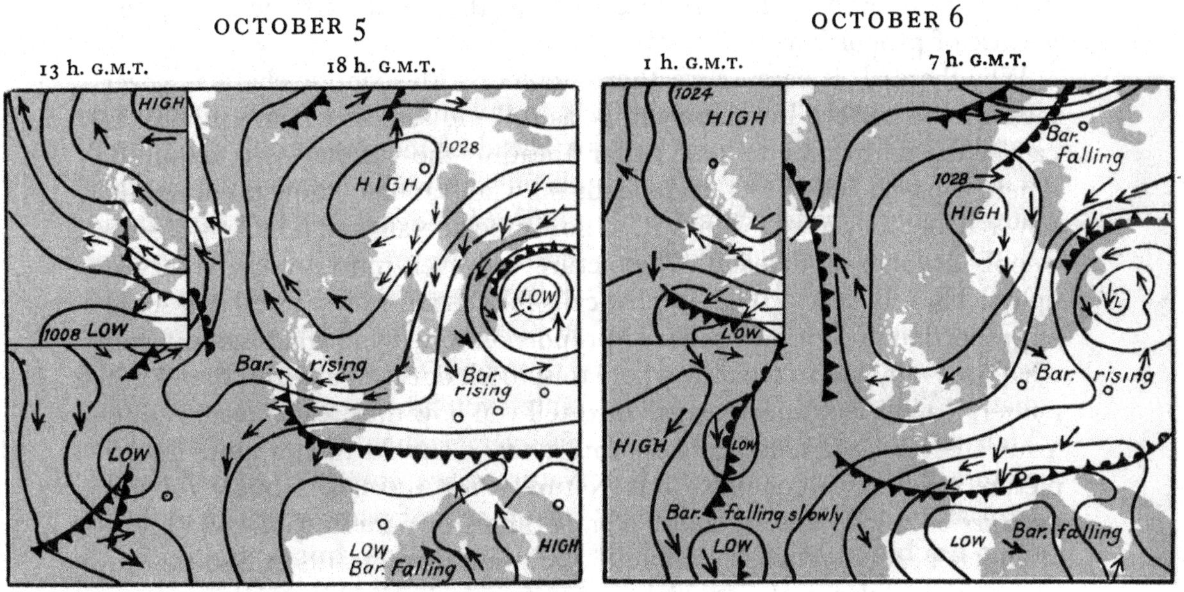

Fig. 26. Anticyclone ('high') over Norway while Baltic depression dies out (cyclolysis) and a Mediterranean 'low' is born (cyclogenesis).

ceiling. Whilst mixing with the subsiding dry air mass, it will help not only to make the cumulus cloud tops spread into a layer but also prevent the layer evaporating at once. When near enough it may indeed make an alto-cumulus layer of its own as we notice at 1050 this morning. That is the characteristic medium cloud type 5. To-night it clears, but this time there is no fog. We are evidently in a drier air mass.

6 and 7 October show much the same air-mass characteristics.

Back to first principles in Chapter 1: 'The best way to use what is known, in order to estimate what has hitherto been unknown, is to plot data on charts and then draw the charts up. In experienced hands this is a re-markably rapid and accurate method, combining science with art.' Our central forecasting offices must have some of the most experienced hands in the country, if not in the world. With their charts as a guide you should be in good hands indeed, even if not perfect. Remember their limitations. First big working charts must be plotted at speed, as fast as the data come in. The plots show all that is known for certain. There may not be many. Even if many, they are only scattered, with areas blank in between. Even if crowded, they may have errors. All else must then be *inferred* by an expert mixture of experience with thermodynamical or hydrodynamical laws as well as many less definite rules of thumb, an artistic sense, and a sense of proportion.

Why these? It is because weather inference with synoptic charts is never meant to be perfect, but merely to be as full and as nearly true a picture as can be drawn in time to keep as far ahead of the weather as is required. There is psychology in it. It depends on who wants to know the weather, how reliably they need to know it, in how much detail, and how long ahead. Supply depends on demand. The picture, however, must always be served up quickly and clearly. That is what calls for artistic sense as well as a sense of proportion. You need a sense of proportion to tell you what matters and what does not, and so to know where to smooth out irregularities that don't matter. Your scientific senses will next tell you how they really average out. Your artistic sense tells you how this average should then appear in the picture. Nature is complex; but Nature *when averaged suitably for our purposes* on a map always shows a smoothness of pattern that to an artist often has beauty. When, therefore, he draws isobars, fronts etc. so that to his artistic eye they look 'right', he has in fact probably 'hit off' the average truth more accurately in his picture than anyone else could.

To forecast the movements of 'highs', 'lows' and fronts, for instance, the three standard methods described at the end of the preceding chapter depend quite obviously on accurate isobar drawing. To be able to 'hit off' the isobars' spacing and curvature with the very first bold quick strokes of a pencil, calling for little or no india-rubber, is an artistic gift that a forecaster should appreciate. Not only is the finished work a thing of beauty, but also it saves much vital time in the first one or two hours when the chart is used for an up-to-the-minute forecast.

When little or no more data are due, the working chart is finally boldly drawn up as the official picture, which is then broadcast. *D.W.R.* charts, for instance, are copied from it and lithographed. Ours in this book, in turn, are copied from *D.W.R.*'s and simplified. Even further simplification will be suggested. For Meteorological Office use, on the other hand, the official picture is simply described in figures, such as latitudes, longitudes and pressures at pressure centres, ends of fronts and isobars. Plotting these points on blank charts and filling the lines in accordingly will reproduce approximately the official picture. Again an artistic hand is an asset.

Weather maps are not reproduced all the way through this book, but enough are included to illustrate all kinds of systems and how they should look when properly charted. No two are ever identical, but at least you will notice the smoothness of curvature of all fronts (except at depression centres), their tendency to be convex forwards, and the smoothness of change of isobars' spacing and curvature in each air mass between fronts. Apart from fronts the only sharp kinks are occasionally to be seen among mountains, where sea-level isobars are of course apt to be meaningless anyway.

No upper-air charts are reproduced. This does not mean that the official pictures were drawn up without them, nor that they can nowadays be so; but by using ground-level observations alone both writer and readers can better appreciate what use they are. From them all our upper-air inferences are drawn. 'Surface' or 'sea level' charts are still the most informative kind, though upper air charts are extensively and regularly used nowadays as well. Upper air data are published by the Meteorological Office in this country in the form of a *Daily Aerological Record*, corresponding to the *Daily Weather Report*.

2. ILLUSTRATIONS OF ANTICYCLONES

	Place no.	Date 1936	Time G.M.T.	Weather letters	Cloud				Vis. V.	Wind		Temp.	
					T.	L.	M.	H.		D.	F.	F.	C.
[a]	01	8.10	0700	c m$_0$	7	4	7	–	5	Calm		44·0	07
	01	8.10	0810	o m d$_0$d$_0$	8	4	2	–	4	N	1	44·0	07
	18	8.10	1100	o m$_0$ r$_0$r$_0$	8	0	2	–	5	NNE	1	47·6	09
	18	8.10	1250	bc	4	1	5	–	6	E	3		
	19	8.10	1315	bc	3	2	5	0	6	E	3		
	01	8.10	1615	c	7	0	5	–	6	E	3		
	01	8.10	1940	c m	7	–	–	–	4	E	2	45·6	08
	01	9.10	0830	c m	7	0	5	–	4	NE	1	45·0	07
	18	9.10	1100	c m u	7	3	–	–	4	ENE	2	52·0	11
	18	9.10	1220	o r$_0$ > R > r	8	3	–	–	5	ENE	2		
	18	9.10	1300	o/r	8	8	–	–	6	ENE	2		
	18	9.10	1630	o	8	0	3	–	6	ENE	2		
	01	9.10	2000	b m$_0$	0	0	0	0	5	ENE	2		
	01	9.10	2200	c	7	0	3	–	6	ENE	2		
	01	10.10	0800	bc m	6	0	3	–	4	NE	2	44·8	07
	01	10.10	0830	o m iR	8	3	–	–	4	NE	2		
	18	10.10	0900	o m$_0$ r	8	3	6	–	5	NE	2		
	18	10.10	1100	bc	3	3	6	–	6	ENE	3	51·5	11
	01	10.10	1510	bc	2	2	6	–	6	NE	3	50·6	10
	19	10.10	1545	c	7	3	6	–	6	NE	3		
	01	10.10	2245	b	1	0	6	0	6	NNE	3	44·2	07

[a] Mist 400–500 ft deep.

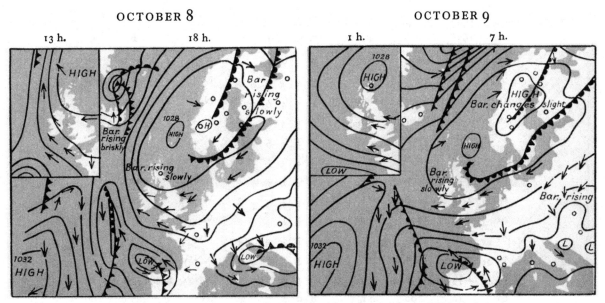

Fig. 27. Anticyclone spreading over Europe with polar continental (P_c) air bringing mainly fair weather from the north-east.

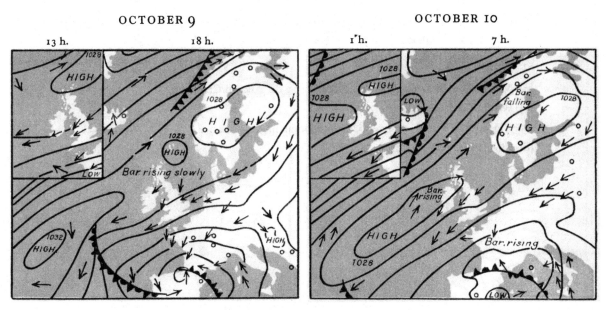

Fig. 28. European and Atlantic anticyclones joining up over Britain.

SYNOPTIC CHARTS

	Place no.	Date 1936	Time G.M.T.	Weather letters	Cloud				Vis. V.	Wind		Temp.	
					T.	L.	M.	H.		D.	F.	F.	C.
	01	11.10	1000	b m	0	0	0	0	4	N	2	47·7	09
	01	11.10	1015	bc m	3	5	0	0	4	N	2	48·1	09
[a]	19	11.10	1230	bc m	5	1	6	0	5	N	2		
	19	11.10	1245	c m U	7	3	6	–	5	N	2		
	01	11.10	1300	c	7	3	6	–	6	N	2		
[b]	01	11.10	1430	c	7	3	6	–	6	NNE	2	52·3	11
	01	11.10	2200	b f	0	0	0	0	2	Calm		43·8	07
	01	12.10	0820	o f	8	0	7	–	3	W	1	45·5	07
	01	12.10	1400	o z_0	8	0	2	–	5	W	1		
[c]	01	12.10	1650	c z	7	3	7	–	4	NW	1	53·5	12
	01	12.10	1715	o z	8	3	7	–	4	NNW	3	53·6	12
	01	12.10	2200	o m	8	–	–	–	4	NNW	2	50·0	10
	01	13.10	0700	o r_0	8	0	2	–	6	SW	3	47·6	09
[d]	01	13.10	0820	c m	7	0	7	–	4	W	3	48·7	09
	01	13.10	0830	c m	6	0	2	8	4	W	2	48·1	09
	18	13.10	1100	bc z_0	6	1	0	8	5	W	3	56·3	13
		13.10	1250	o z_0	8	0	2	–	5	WNW	2		
	01	13.10	1600	o z_0	8	0	7	–	5	NW	2	52·4	11
	01	13.10	2200	b m_0	0	0	0	0	5	NW	1	48·8	09

[a] Mist or haze of London smoke, making very dark sky in Cb cloud.

[b] Weather becoming fair.

[c] Gloomy sky with thick haze. C_M7 predominating.

[d] Clouds (C_M7) moving from NNW.

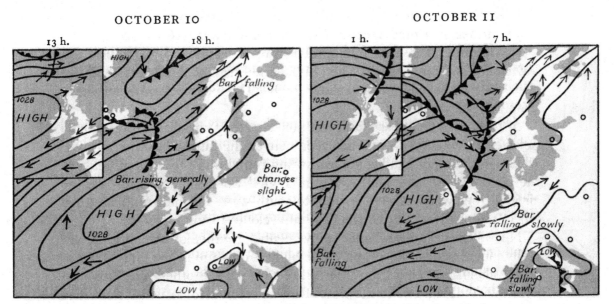

Fig. 29. Shift of high-pressure centre from Europe to the Atlantic, with change
from easterlies to westerlies over Britain.

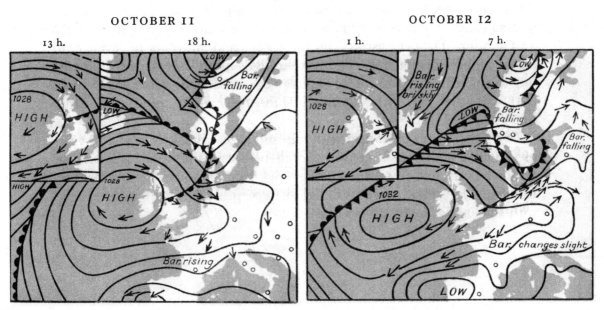

Fig. 30. Main anticyclone now over Atlantic, bringing maritime (M) air with
cloudy weather from the north-west.

8 October is at first sight similar in main features to 5 October.

9 October is at first sight similar in main features to 6 October.

10 October is at first sight similar in main features to 7 October with low clouds and rain as well. The only charted feature that looks at all like their cause is the old occlusion shown on 8–9 October on its way south-west from the Skagerrak.

Of this broad and nearly uniform north-easterly air stream, presumably polar continental (P_C), some has been converging into the southern 'lows' whilst on the other hand some has swung round into the 'col' in between the north-eastern and south-western 'highs' where it meets the Atlantic air at a front which it sweeps away to south-west. No old fronts are then left in the picture, but just the two anticyclones fusing together.

This illustrates anticyclones' movements. Neither of the anticyclones really moves bodily like a 'low'. The highest pressure is just quietly transferred along the ridge from one centre towards the other. We finish up with the dominant one to the west of us, all set now for Atlantic westerlies to come to us round it behind a warm front.

The 'low' or trough developing on the point of occlusion between Iceland and Norway runs freely south-east. Our own end of the warm front, of course, being so much nearer the high-pressure centre, is slower and so gives only a trace of rain but much concentrated smoke haze on **12 October.** The overcast sky, on the other hand, prevents recurrence of actual night radiation fog. The next trough, meanwhile, a wave on the long front still linked to the previous one, comes along more freely than ever with the 'bar.' (i.e. barometric pressure) falling ahead and rising briskly behind it.

At 07 G.M.T. on **13 October** with rain and haze, frontal clouds and a veer of wind, we are to see it arrive. Clouds persist as the front trails almost stationary off Land's End and another cold front does likewise on reaching London. The 'ridge' keeps us in light winds and haze as we should expect with such a big 'high' still so near.

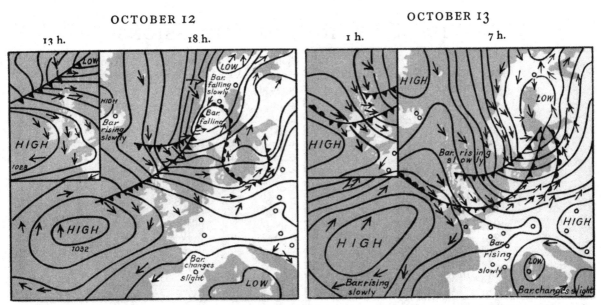

Fig. 31. Trough moving south-east with sector of westerlies (maritime, M air, warm, cloudy) followed by north-north-westerlies (Arctic, A air, cold, showery). Note how it develops on the lee side of the mountains of Norway, where it occludes.

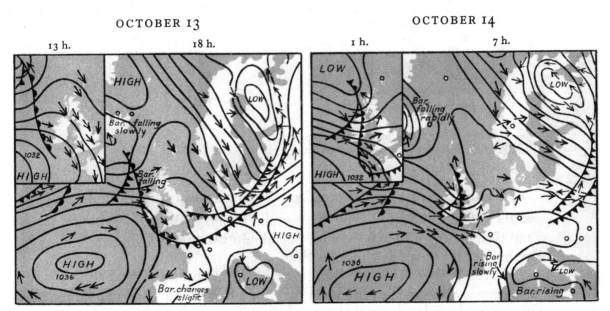

Fig. 32. Second trough moving south-east over Britain, showing warm front and starting occlusion.

3. ILLUSTRATIONS OF DEPRESSIONS

	Place no.	Date 1936	Time G.M.T.	Weather letters	Cloud				Vis. V.	Wind		Temp.	
					T.	L.	M.	H.		D.	F.	F.	C.
	01	14.10	0820	o	8	0	2	–	6	SSW	3	50·9	11
	18	14.10	0900	o rr	8	7	2	–	6	SW	3		
	01	14.10	1300	o	8	5	2	–	6	WSW	3	55·0	13
	01	14.10	1600	o m	8	0	2	7	4	W	3	56·1	13
	01	14.10	1800	o m rr	8	7	–	–	4	W	3	56·1	13
	01	14.10	2200	c	7	0	3	–	6	WNW	3	54·5	13
	01	15.10	0730	c	7	0	7	–	6	W'S	3	52·5	11
[a]	01	15.10	0800	bc	5	0	0	6	6	W'S	3		
	18	15.10	1100	c	6	1	0	6	6	WSW	4	64·0	18
	01	15.10	1630	c	7	4	0	2	6	WSW	3	61·0	16
	01	15.10	1900	bc	6	0	0	2	6	WSW	3	55·9	13
	01	16.10	0720	bc	2	0	0	9	6	W'S	3		
	18	16.10	1100	bc	4	1	0	2	6	W	4	58·0	14
	18	16.10	1230	c	7	4	0	2	6	W	5		
	01	16.10	1800	bc	4	4	0	2	6	W'S	3		
	01	16.10	2100	c	6	5	0	2	6	W'S	4	52·6	11
	01	17.10	0700	c	7	5	–	–	7	W	6	53·5	12
	20	17.10	1700	c	6	5	0	6	7	W	5		
	20	17.10	2130	o rr	8	–	2	–	7	W	5		
	01	17.10	2240	o d	8	5	2	–	9	W	6	55·8	13
	01	18.10	1015	b q	1	1	0	0	7	W	5	55·0	13
[b]	01	18.10	1430	bc q_0	3	2	0	0	7	W	5	56·4	14
	01	18.10	2230	c	7	0	0	6	6	W	2	46·8	08
	01	19.10	0730	o	8	7	2	–	7	W'S	6	48·0	09
	01	19.10	0830	c rr	7	7	2	–	7	W'S	6	50·0	10
	18	19.10	0930	bc	5	3	7	0	7	W	5		
	18	19.10	1100	bc	3	2	0	0	7	W'N	6	56·5	14
	18	19.10	1300	o d	8	8	0	0	7	W'N	4		
	18	19.10	1700	o pr q_0	8	8	–	–	6	NW'W	5	51·3	11
	01	19.10	2200	b	1	0	6	0	7	NW	5	45·1	07

[a] $C_H 6$ in peculiar forms, including enormous tufts.
[b] Weather becoming fine, sky cloudless from 17 to 22 G.M.T.

14 October. A new wave depression comes in, showing first the normal warm frontal features with continuous rain, then the wind-veer with mist, and then the shorter period of cold frontal rain with further wind veer and clearance to broken altocumulus clouds.

The charts illustrate how the wave occludes, and how a separate centre of low pressure tends to be formed on the point of occlusion. Breaking off its long trough from the Iceland 'low' in the afternoon it finally makes it.

15 October. This does the same thing on a bigger scale except for being already occluded before it arrives here. The only clear sign of it here, in

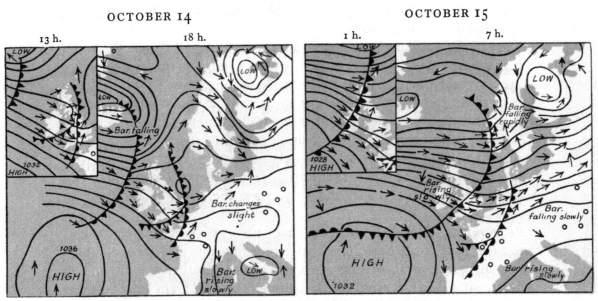

Fig. 33. Trough (rainy), with occluding warm sector (T_m air, warm, cloudy) followed by another occlusion bringing west-north-westerlies (P_m air, cool, fair).

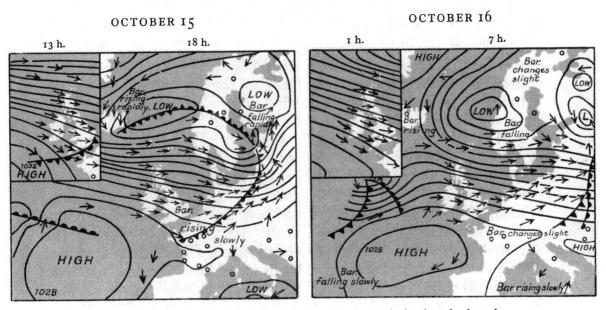

Fig. 34. Westerlies (P_m air, cool, fair), also showing occlusion bent back as the Arctic front between cool P_m and cold A air.

fact, is the peculiar form of high cloud which presumably is the remnant of a warm sector now high aloft.

Then comes a striking specimen of a straight stream of P_m air. With bar. seen to rise to 1020 mb. we are evidently close enough to the high-pressure centre—or rather the high pressure comes fast enough to us—for subsidence aloft to make the cumulus clouds soon spread out and vanish, whereas farther north they are able to grow into cumulonimbus with showers.

Also the charts show how the occlusion trails back, not only south-west through the 'high' as a link with the next main wave, but also north-west in a long trough where it acts as the arctic front between cool P_m and cold A air. There it is bent back, and so you see the effect of a BACK-BENT OCCLUSION.

16 October. Ahead of the next main wave we keep similar 'ridge' conditions with small cumulus clouds spreading out into stratocumulus low cloud type 4, until evening when low clouds regather as stratus. The charts illustrate the ridge and the col between the two 'highs' and between the old and new 'lows'. They also show a suspected wave just beginning to form on the trailing cold front.

17 October. Low clouds of stratus type, apparently due to the moisture and turbulence of the strong maritime wind, to-day obscure all frontal clouds until evening when rain begins. The rain, however, does not seem to last long, and from the chart we see that not only are we still near to the 'high' but also the 'low' itself is mostly occluded. Having been deepening all the morning (Iceland bar., for instance, falling, yet with the low-pressure centre already past), it is now fully developed. At 07 G.M.T. on 16 October its centre appears near a certain ship (*Scythia*) at about 52 N 34 W. At 18 G.M.T. on 17 October, just 35 hr. later, it is at 61 N 01 W. So it has moved about 1300 miles or 2000 km. east-north-east in 35 hr. with mean velocity therefore about 37 m.p.h., 16 m. per sec. or 32 knots from west-south-west. Compare with the mean warm-sector gradient wind, 40 knots, also from west-south-west, and you see how nearly the WARM-SECTOR GEOSTROPHIC OR GRADIENT WIND can be used to predict the centre's movement. Occluded, however, without a warm sector at all, the depression may now slow down. Already we see its motion turn from east-north-east to due east. Two wind arrows are shown near it on the morning chart (07 G.M.T.) for 16 October. They are respectively from north-west (a gale with temperature 42° F. reported by ship *Scythia*), and south-south-west (a gale with temperature 62° F. reported by *Queen Mary*). Naturally the depression's cold front is assumed between them.

OCTOBER 16

13 h.   18 h.

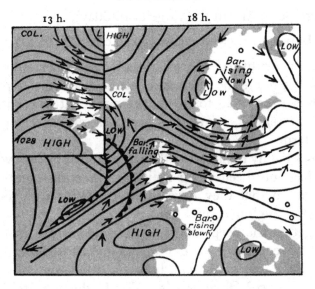

OCTOBER 17

1 h.   7 h

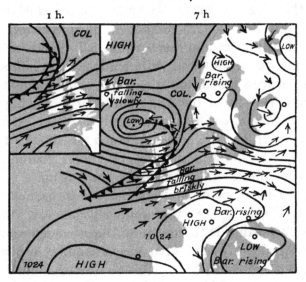

Fig. 35. Ridge (cool, fair), followed by depression deepening and occluding (rainy), also showing col at first over Iceland, moving east.

Queen Mary, steaming *westward*, has reported *barometric tendencies*

+3·8 mb. per 3 hr. (in the ridge at midday),

+0·6 mb. per 3 hr. (when passing the crest of the ridge at midnight),

and −6·0 mb. per 3 hr. (in the warm sector next morning).

Now in the 35 hours the depression centre itself has *deepened* from 999 to 964 which is 35 mb. So to the total tendency it has contributed 1 mb. per hr. fall of pressure=−3 mb. per 3 hr. That may therefore be called the 'deepening' term.

Advection or bodily transport of the warm-sector isobars with the centre, meanwhile, has only been along their own direction west-south-west to east-north-east, so that they have in effect stayed still and contributed nothing. That is the 'advective' term.

Adding together the deepening and advective terms −3+0, we get the total barometric tendency, −3, due to changes of pressure lines relatively to the earth. *Relatively to the ship*, however, it is −6.

Why the difference? Obviously it is due to the ship itself, relatively to the earth, so crossing the isobars as to account for 3 mb. in 3 hr. Steaming westward on this occasion at about 30 knots, the *Queen Mary* covers within 3 hr. the whole 90 mile westward component of gap between 3 mb. isobars.

Scythia, meanwhile steaming eastward at half of this rate in 3 hr., has covered 1–2 mb. isobar interval with tendency 1–2 units, implying no tendency left between the air and the earth. Sure enough, at noon next day, when finding itself in the same part of the moving depression as *Scythia* was, the *Stavangerfjord* at rest relatively to the earth has tendency zero. Six hours later, however, when in much the same part of the moving depression themselves, the Faroe Islands report barometric tendency nearly +6. The depression has started to fill up again.

Our next polar-front wave (No. 6 of this series) has not developed like those before it, of which no two were quite alike anyway. As there is more of a 'high' on its north side to hold it up, the scene appears set for arctic air now to sweep down freely. That means trouble, probably short and sharp. We have to look out for disturbances from the north-west.

18 October. The first thing for which to look out is always a back-bent occlusion. A short one appears on the midnight chart, but you see there has not been a long trough in which any serious length of occlusion can ever have trailed to become bent back; so we look instead for the *non-*frontal trough that is apt to veer the wind with squalls and showers (or rather, inland, a single but long shower) behind a depression of this sort.

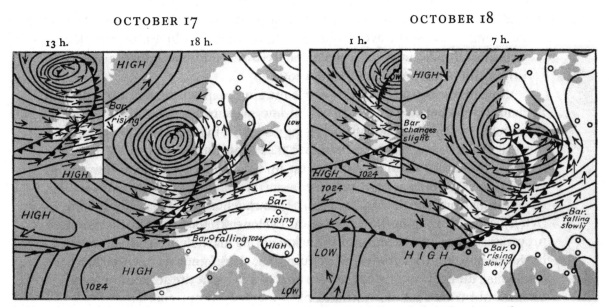

Fig. 36. Depression or cyclone (stormy) at its deepest, fully occluded, showing motion.

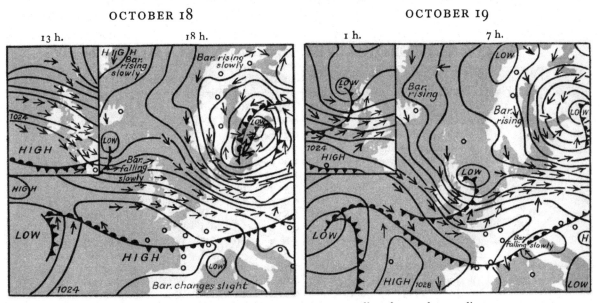

Fig. 37. Polar 'low' bringing cold north-westerlies after cool westerlies.

It fails to appear distinctly here, but is quite pronounced over Denmark this evening. This still leaves us in cool P_m but not in cold arctic air. Our cumulus clouds are small, the wind is westerly, and the temperature is still up in the fifties Fahrenheit, above 10° C.

Something turns up at last on the evening map. A suspicious trough to Iceland has deepened into an actual low-pressure centre there. With bar. reported to be falling ahead and rising behind it in such a picture, we should forecast it here next day. Probably it is a POLAR DEPRESSION. That is the initially NON-FRONTAL circulation set up by sufficiently general heavy convection in cold upper air on warm sea. The air expands upward, you see, and thus contracts horizontal area (A) so much as to increase cyclonic spin (R). That is roughly how a tropical cyclone works too, though the exact processes are by no means fully known. This new depression of ours, however, appears with fronts. If genuine, therefore, they have either been formed on the spot or else they are part of a wave in the arctic front or some such old front that has not hitherto been marked on our charts.

19 October's picture looks at first rather like 14 October's; but our main front this time cannot be the same polar front, for the latter by now has been traced farther south. It looks more like the old arctic front.

4. COLD FRONT DEVELOPMENT AND MOTION

Place no.	Date 1936	Time G.M.T.	Weather letters	Cloud				Vis. V.	Wind		Temp.	
				T.	L.	M.	H.		D.	F.	F.	C.
01	20.10	0740	b	1	0	0	1	6	NW	3	44·7	07
01	20.10	0800	bc	2	0	0	4	6	NW	3	46·1	08
18	20.10	1100	bc	2	0	0	4	7	NW	4	51·0	11
18	20.10	1300	bc	4	0	0	5	6	NW	3	52·9	12
18	20.10	1445	bc	4	1	6	6	6	NW	3		
18	20.10	1530	bc	5	3	6	6	6	NW	3		
18	20.10	1715	o	8	0	7	6	6	NW	3		
01	21.10	0730	o	8	0	2	–	6	W	3	50·0	10
18	21.10	1145	o	8	0	2	–	6	W	3		
18	21.10	1430	o	8	5	2	–	6	W	3	55·3	13
01	21.10	1900	bc	6	5	–	–	6	W	3	52·7	11
01	22.10	0730	c	7	0	3	2	6	W	3	48·6	09
18	22.10	1230	c	7	4	3	0	6	W	2	63·5	17
01	22.10	1715	c m	7	0	3	0	4	W	2	56·5	14
01	22.10	2145	o m$_0$	8	–	–	–	5	W	1	55·3	13
01	23.10	0720	o/r	8	0	2	–	6	SSW	3	53·0	12
18	23.10	1100	c	7	4	0	–	6	SW	3	57·0	14
18	23.10	1340	o/d$_0$	8	9	2	–	6	SW	3		
01	23.10	1620	c/d	7	0	3	–	6	SSW	4	54·3	12
01	23.10	1930	bc	5	0	3	–	6	SSW	3	51·7	11

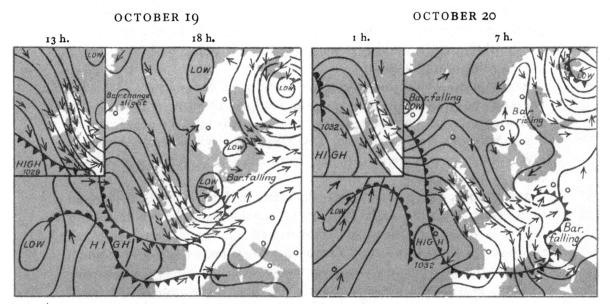

OCTOBER 19

13 h.   18 h.

OCTOBER 20

1 h.   7 h.

Fig. 38. North-westerlies (P air), cold and showery when curved round the 'low', but
warmer and fair where curved round the 'high'.

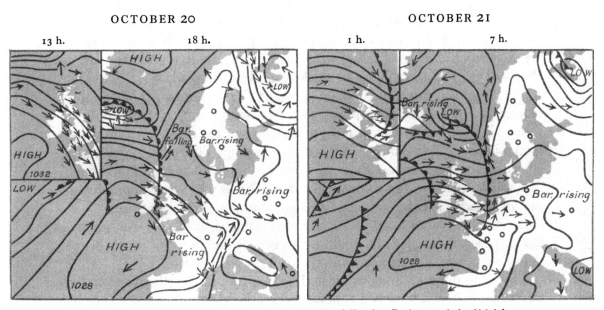

OCTOBER 20

13 h.   18 h.

OCTOBER 21

1 h.   7 h.

Fig. 39. Westerlies, M air, warm and cloudy, following P air round the 'high'.

The occluding frontal belt having passed us by 0930 on 19 October, polar air follows it with much cumulus, clearing at night as usual. Being presumably arctic air which is warming up, it is doing its best to bring a good ridge of high pressure.

20 October. The pictures resemble those of 10 days earlier. The ridge extends and strengthens the main anticyclone to west, and again the Atlantic westerlies start approaching us round it behind a warm front.

21 October. Neither this nor anything in the warm sector, however, gives worse weather here than just an overcast sky. Again you see the effect of high pressure.

22 October. The picture now really starts to differ from its 10-day-old predecessor. With the 'high' further east now blocking the way instead of helping it round, the Iceland 'low' moves not south-east but north-east, and so we stay in the warm sector. Temperature reaches the sixties Fahrenheit and with the high-pressure centre so near us we see only a few low clouds.

The charts show a big cold front. Note how the isobars in the warm sector ahead are nearly straight, making the true wind geostrophic, whereas those behind are curved round the anticyclone in the south-west so that their true or gradient wind is stronger than geostrophic. The result, of course, is that the true winds converge horizontally at the front there, helping FRONTOGENESIS or intensification of the front. Northward, however, this effect obviously decreases until, where isobars behind the front are no longer anticyclonically but cyclonically curved (round the 'low'), they will help to FRONTOLYSE or weaken the front instead.

The front's geostrophic rate of advance (V_{fgs}), remember, is simply marked by the spacing of points where neighbouring isobars cross it. Except in the north to-day you will notice this spacing quite wide, implying light winds and slow motion. A wave may develop. That would completely alter the local isobars so as to spoil our simple calculation. So much depends on so little. Instability is like that. That is why weather prediction cannot be exact like the Nautical Almanac. Sun, moon and stars are stable. Our air, on the other hand, being sometimes unstable, even if ever known just as precisely as they are, could never be forecast in detail much better than now. Slightly better, admittedly, but not very much.

That is important. In somewhat the same way as the discovery of the Uncertainty Principle, only a generation ago, for the first time showed the very idea of any perfectly certain prediction whatever (however long taken for granted) to be an illusion, so is the quest of anything *near* perfection in weather prediction a mere wild-goose chase. You may not have realised

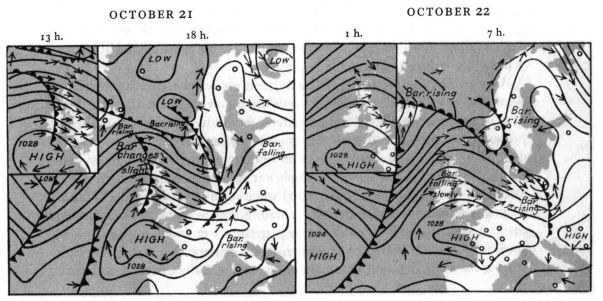

Fig. 40. Sector of M air moving slowly east with the 'high', followed by cold front.

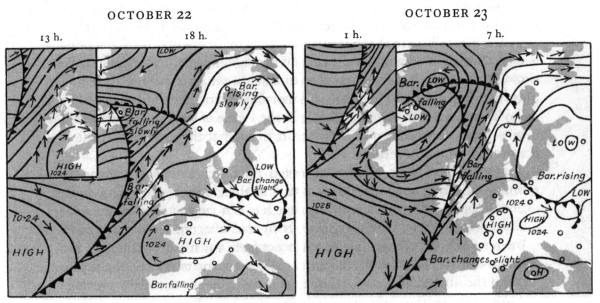

Fig. 41. Cold front, followed by isobars more curved round the 'high' in the south than in the north, causing more convergence and rain.

that. It is all right when stable. Even *waves* can be forecast when stable. But of unstable things, as of whirlpools over a bath plughole or even of the mere fall of anything top-heavy, no details can be predicted at all until you have seen something actually begin.

A slow-moving front must therefore be carefully watched for the slightest sign of a wave. To-day's, however, develops no wave, so it duly arrives next morning as we might have expected. After all, with definite main high pressure to south and low to north creating an eastward push, whether disturbed or not, the front would be bound to get here somehow. That would be certain. But when? The slower it is, the more uncertain is its estimated time of arrival (abbreviated to e.t.a.) a long way ahead. On the map as a whole you could not go far wrong, but to any one place it might easily make all the difference between getting wet and not getting wet—success or disaster, whether to garden-party or harvest. That is how forecasting for a single place, at first sight so easy, is really often most difficult.

5. DEPRESSION DEVELOPMENT AND MOTION

24 October. The front having passed, the light wind immediately behind it has allowed fog to be formed in the cool air after the rain. Rain does not appear here in the record, but 0·11 in. of it was observed in a rain-gauge which was empty the previous morning. As the afternoon drizzle could not possibly have amounted to all this, we must infer that it rained quite a lot in the night. The fog this morning clears quickly into an ordinary day of fair-weather cumulus clouds in the P_m air of the high-pressure ridge until evening, when upper clouds start to suggest the approach of the first warm front of a new series.

25 October. This is next observed in vigorous action at 0630 next morning. All over soon after 07 G.M.T. as the chart shows, again we have normal polar-air weather, but this time less of a ridge and so more convection to cause a heavy thundery shower by mid-afternoon. After the shower the wind falls light, though apparently not from any marked widening of the isobar spacing. The local big storm-cloud's air currents may be the cause. Then the air becomes cooler, misty and more stable as usual by evening.

Meanwhile in the Atlantic the liner *Montrose* steaming westward at about 54 N 35 W has reported successively:

(i) fresh south-east wind with drizzle, temperature 46° F., and barometric tendency falling 8 mb. in 3 hr.;

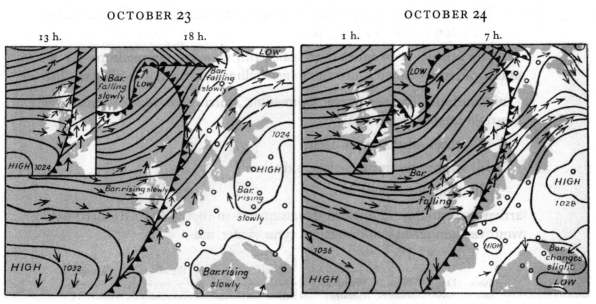

Fig. 42. Cold front followed by isobars wider apart in the south than in the
north, causing lighter winds.

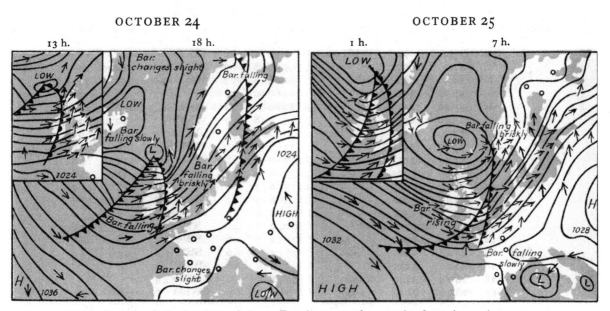

Fig. 43. Orthodox type of chart. Development of new polar-front depression.

(ii) strong west-south-west wind with temperature 53° F. and barometric tendency −6 mb. per 3 hr.;

(iii) squally west-north-west gale with showers, temperature 44° F. and barometric tendency now no longer falling but rising +6 mb. per 3 hr.

26 October. The next wave depression, which these observations have clearly located respectively:

(i) ahead of the warm front,

(ii) within the warm sector, and then

(iii) behind the cold front,

arrives here to-day. The warm front appears to be through by 1250 with wind veer and change to low-cloud type 5 after rain.

	Place no.	Date 1936	Time G.M.T.	Weather letters	Cloud				Vis. V.	Wind		Temp	
					T.	L.	M.	H.		D.	F.	F.	C.
	01	24.10	0720	c F	7	5	–	–	1	SW	2		
	01	24.10	0730	bc m	5	5	0	9	4	SW	2	48·7	09
	18	24.10	1145	bc	2	5	0	0	6	WSW	3	49·5	10
	01	24.10	1400	bc	2	1	0	0	7	SW'W	4	56·5	14
	01	24.10	1540	b	1	4	0	0	6	SW'W	4	54·0	12
	01	24.10	1930	bc	5	0	0	2	6	SW	3	49·7	10
	01	24.10	2200	bc q_0	2	0	5	2	7	SW	4	49·8	10
	01	25.10	0630	o RR q	8	7	–	–	6	WSW	6		
	01	25.10	0950	c	7	3	9	8	6	W'N	6	52·1	11
	01	25.10	1045	b	1	1	0	8	7	W'N	6	54·1	12
	01	25.10	1310	b	1	1	0	0	7	W	5	53·3	12
	01	25.10	1410	c	6	2	0	6	7	WSW	5	52·8	12
	01	25.10	1530	c U t_0	7	3	0	8	7	WNW	5	50·2	10
	01	25.10	1540	c u PR	7	3	0	8	6	WNW	7	47·5	09
	01	25.10	1545	c p r	7	3	–	–	6	WNW	4	46·9	08
	01	25.10	1555	c r_0	7	3	–	–	6	W'N	3	47·1	08
	01	25.10	1605	c m_0	7	3	6	–	5	W	2	47·0	08
	01	25.10	1635	bc	4	3	0	3	6	W	2	45·8	08
	01	25.10	1650	bc m_0	3	3	6	3	5	W	3	46·0	08
	01	25.10	2150	b	1	0	6	0	7	W	3		
	01	26.10	0730	o/r	8	0	2	–	6	S'W	3	46·1	08
	01	26.10	0740	o rr q_0	8	0	2	–	6	SSW	4	46·2	08
[a]	18	26.10	1100	o r_0r_0	8	7	2	–	6	SSW	6	51·9	11
	18	26.10	1250	o	8	5	2	–	6	SW	4		
	18	26.10	1340	o r_0r_0	8	7	2	–	6	WSW	4		
	01	26.10	1620	o r_0r_0	8	7	2	–	6	WSW	6	56·8	14
	01	26.10	1830	c $r_0>$i R	7	7	2	–	6	W'S	5	56·2	13
	01	26.10	2030	bc	4	7	7	–	7	W'S	6	55·3	13
	01	26.10	2145	c	7	0	3	–	7	W'S	6	54·7	13
	01	26.10	2220	o	8	0	2	–	8	W'S	7		

[a] Rain ceases about noon.

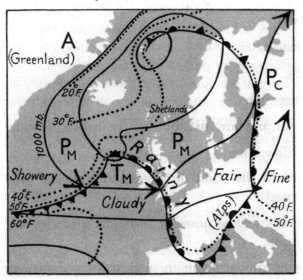

Fig. 44. Simplified chart of PRESSURE (sea-level), marked by isobars at intervals of 16 mb. from 1000 mb., showing: AIR-FLOW (low level), marked by arrowheads on the 1000 mb. isobar; W.P.T. (estimated), marked by isotherms at intervals of 10° F. from 0° F. or approximately 5° C. from 0° C.; AIR-MASS TYPES, marked by initials (e.g. A for arctic); FRONTS, marked by standard symbols; and WEATHER TYPES.

This chart shows not only the usual features but also the weather itself, and how simply all are determined by the air-flow and the potential wet or equivalent temperature distribution. You may also notice how mountains such as the Alps affect them.

This morning's chart is the last to be reproduced on the small scale in this book. But it is not like the others. It is a specimen of a simpler type of weather map showing:

(i) Potential wet or equivalent temperature distribution (by isotherms).

(ii) Air masses.

(iii) Fronts.

(iv) Weather types.

(v) Sea-level pressure distribution (by isobars at wide intervals).

(vi) Wind-flow lines,

(a) drawn with the wind directions, and spaced apart inversely as wind speed;

(b) spaced as widely as possible, provided that they show all the main air streams;

(c) always meeting at fronts in such ways as air streams must;

(d) each to be traced if possible right across the map;

(e) always clearly accounting for all the air.

Of course (ii)–(iv) can be inferred from (i) and from one another, subject to (v) and (vi); while (v) and (vi) are identical or at least have to be consistent with each other.

Consider those principles carefully and you will see how remarkably easy it is to draw up a chart of this kind quickly from scanty data. The same technique can even be used in the tropics, even though the technique of forecasting with it is rather different. From (i) to (iv), by the way, the difference is less than you might expect. Only (v) and (vi) really differ enough to call for separate drawing. Wind-flow lines are based then more upon observations of actual wind than of pressure alone. Again an artistic hand is an asset in drawing them up correctly and quickly. To make the charts really useful at all, in fact, it is almost essential. When gradient wind, depending simply on pressure known accurately, is nevertheless too far from the truth, the true wind has to be taken upon its own merits and forecast accordingly more from the face of the charts alone than from the cast-iron basis of hydrodynamics. Finally this kind of weather map is suggested for showing clearly to air crews and laymen not only the past but also the forecast weather.

Our warm front's arrival here at midday is exactly confirmed by the official midday charts, although they are not reproduced in this book. You will find we can manage quite well now just with descriptions of the original charts, or rather of just the features that matter to us. To follow the special chart for this morning there comes the midday chart showing our warm front clearly marked by the change of temperature, wind, barometric tendency, clouds and rain. From the centre of the depression this front is linked to yesterday's which is now over Europe from the Gulf of Genoa (Mediterranean) to the Arctic Ocean. Two arctic stations provide a fix. One of them is Bear Island, where an east-north-east gale with snow is replaced by a south-west breeze with rain. That shows the front all right. The other is Jan Mayen, where with heavy rain and a north-easterly gale the bar. has fallen to 964 mb. That is so low as to be evidently about the old low-pressure centre. It is due north from the Shetland Isles area where we remember seeing it on the evening of 24 October; so you see how, after complete occlusion, the whole depression changes its motion.

The new depression centre already shown by *Montrose* at midnight 25–26 October to have been about 54 N 35 W, is next revealed at midnight 26–27 October over the Shetland Isles where bar. reaches minimum of 960 mb., while the wind with force 6 or 7 (on our Beaufort scale) backs right round from south to northerly with some snow. Those two positions being 1200 miles apart, the centre's velocity has been 50 m.p.h. (from west-south-west). It now goes on eastward (parallel to the warm-sector isobars, as it normally should be) until over Sweden, from which it turns north-ward to merge itself into the old arctic low-pressure centre (shown by contrasting Spitzbergen's north-east to easterly wind, Bear Island's south-westerly and Jan Mayen's north-easterly becoming north-westerly wind. With those alone plotted upon a chart we at once see the circulation and hence its approximate centre).

The warm-sector gradient wind velocity is not only *parallel* to the centre's motion but has on this occasion just the same average speed (50 m.p.h.) too. Again you see how useful a rule for prediction that is.

Our London surface wind then is south-west to west-south-west with force 6, which is 22–27 knots or 25–31 m.p.h., while ships and exposed west coastal stations like Valentia and the Scillies in the same warm sector themselves have wind force 7, which is 28–33 knots or 32–39 m.p.h. So you see how the inland and sea 'surface' wind speeds (at an effective height of 10 m., anyway) compare with the gradient wind, so that we might estimate one from the other. With lighter winds or stabler air, however, the contrast is greater, the inland surface wind being a smaller fraction of the sea surface wind, and both being a smaller fraction of gradient wind.

This, of course, is a friction effect. Just as an air mass, popularly defined according to *weather*, cannot at the same time be scientifically well defined by temperature alone, nor by moisture alone, nor even by their lapse rates allowing for height, but only by a certain function of all of them (called potential wet or equivalent temperature), so can an air mass according to stability be defined not by temperature alone nor by wind speed alone, nor even by their lapse rates (temperature lapse rate and wind shear with height), but only by a certain function of all of them (known as RICHARDSON NUMBER, or something equivalent). Light winds, small shear or large inversion of temperature will all help to make this number high, and turbulence low. Low turbulence means little disturbance of either the wind or the temperature inversion, which therefore tend to stay as they are. Any tendency for things to stay as they are is a form of stability. Of the three ways of travel of heat, you remember, the first was described as *convection and/or turbulent stirring of the air by friction*, 'the quickest process but conditional on instability or rough flow'. Of stability or instability in the sense of free convection alone we already know much, whereas of stability in the more general sense, including forced stirring by friction or 'drag' on the ground, we have only said little. We now see how it is marked by high Richardson number.

Our warm-sector air to-day, for instance, being ex-tropical air which must have been cooled all the time from below, will accordingly tend to have an inversion to raise its Richardson number and lower the turbulence, so that it is not too well stirred but rather dragged back by the land or rough-sea surface. That makes the surface wind speed not too high a fraction of the undisturbed gradient wind speed. On the other hand, it is forcibly stirred by the very strength of the gradient wind; so its stability can only be moderate.

Meanwhile as *Montrose* bar. minimum was 992 mb. whereas the Shetland Isles report 960, the 'low' has deepened 32 mb. in the 24 hr. That makes average tendency 'deepening term' -4 mb. per 3 hr., with warm-sector 'advective term' again zero, making the total relatively to the earth $-4+0=-4$, whereas relatively to *Montrose* in warm sector it was -6. The difference of 2 mb. is that of the isobars over the 45 miles of track (due west) covered by the ship in 3 hr.

Over England, however, the warm-sector tendency in the south is zero, whilst average tendency ahead of warm front is 1 mb. per hr. Respective east to west components of pressure gradient being zero and about 1 mb. per 25 miles, we deduce by the 'tendency' method

$$V_f = 1 \text{ mb. per hr. divided by } 1 \text{ mb. per } 25 \text{ miles}$$

$$= 25 \text{ m.p.h.}$$

while by the 'geostrophic' method

$$V_{fgs} = 50 \text{ m.p.h.};$$

and then by the 'path' method, using the following plotted positions, namely Pembroke (St Ann's Head), London and Amsterdam, which are at intervals of 210 miles each in 6 hr., we deduce speed 35 m.p.h. with acceleration zero. This illustrates how the three methods compare.

Averaging for V_f in the 'tendency' method is always tricky, and might have been done with more care. Lines drawn through places with equal barometric tendencies are helpful. We call them ISALLOBARS. They, too, of course, have a gradient shown by their spacing. That is accordingly called isallobaric gradient. Places of most rapid rise or least rapid fall of pressure are isallobaric highs, while those of least rapid rise or most rapid fall of pressure are isallobaric lows.

V_{fgs} is a fair enough estimate as such; but no one expects it normally to be the true speed V_f of the front. The front, you remember, must move with the cold-air mass, whose velocity differs from geostrophic owing to

(i) *isobar curvature*, adding a backward component of wind if cyclonic, but forward if anticyclonic;

(ii) *friction*, which slows the wind down and deflects it across the isobars towards lower pressure most noticeably at ground- or sea-level; together with

(iii) *rising or falling pressure*, which pushes the air along the barometric tendency gradient from isallobaric high to low.

Then the true V_f by virtue of (i) must be less than V_{fgs} wherever the isobars are more cyclonically curved in the colder air, whilst by virtue of (ii) it is less anyway near the ground, although by continual mixture with air above it the front has effective speed rather faster than surface wind.

125

With (i), (ii) and (iii) together, in fact, we normally find

$$\frac{V_f}{V_{fgs}} = 60\text{–}80\% \text{ for warm fronts,}$$

$$= 70\text{–}90\% \text{ for cold fronts followed by isobars curved round a 'low',}$$

$$= 100\% \text{ for cold fronts followed by isobars curved round a 'high'.}$$

For to-day's warm front it is about

$$\frac{35}{50} = 70\%.$$

35 m.p.h. is about the same as the sea-surface wind speed but rather more than our land-surface wind speed, as we should expect.

6. CYCLOGENESIS, RIDGE AND WAVE DEVELOPMENTS

27 October. The liner *Ascania*, steaming west about 53 N 24 W at midnight, has reported successively

(i) in the high-pressure ridge: bar. 1021, rising 0·8 mb. per 3 hr.; temperature 50° F.; wind west by north; sky half clouded with small cumulus but also completely covered by increasing alto-stratus clouds;

(ii) at midnight a few hours later: bar. 1011, falling 4 mb. per 3 hr.; temperature 53° F.; wind south-east 5; sky overcast with ragged low clouds of bad weather, and drizzle;

(iii) next morning: bar. 1005, falling 1·4 mb. per 3 hr.; temperature 56° F.; wind south-south-west 6; sky continuing overcast with $C_L 7$ and drizzle.

The warm front clearly located by these observations extends like the last one from its depression centre (south-west of Iceland) to the occlusion of yesterday's system now over Europe from the Gulf of Genoa to Sweden. From yesterday's picture, in fact, we only notice the following differences:

(i) the new centre not coming here but going off to Iceland;

(ii) its warm front therefore coming to us much more slowly than the previous one;

(iii) cyclogenesis as the old cold front is bent round the Alps into the warm wet Gulf of Genoa.

Place no.	Date 1936	Time G.M.T.	Weather letters	Cloud				Vis. V.	Wind		Temp.	
				T.	L.	M.	H.		D.	F.	F.	C.
01	27.10	0730	b	0	0	0	0	7	WSW	6	46·0	08
18	27.10	0915	bc	2	0	0	5	7	WSW	5		
18	27.10	1100	o	8	0	0	7	7	W'S	6	51·0	11
18	27.10	1315	c	7	2	0	8	7	WSW	6	51·0	11
18	27.10	1400	bc	2	1	0	8	7	WSW	6		
18	27.10	1445	bc	3	1	5	0	7	WSW	5		
18	27.10	1530	o r_0	8	1	7	–	6	WSW	4		
18	27.10	1615	o rr	8	0	2	–	6	WSW	5		
01	27.10	1815	bc	2	0	5	8	6	W'N	5	45·7	08
01	27.10	1830	o RR	8	7	–	–	6	W	4	46·1	08
01	27.10	1940	c	7	0	7	–	6	WSW	3	44·0	07
01	27.10	2140	b	2	5	0	0	7	WSW	4	43·0	06
01	28.10	0730	b	0	0	0	0	7	W	3	42·9	06
18	28.10	1000	b	1	1	0	5	7	WNW	5		
18	28.10	1100	c	7	1	0	6	7	NW'W	5	51·0	11
18	28.10	1145	o	8	5	0	7	7	NW'W	5		
18	28.10	1300	c	6	1	0	8	7	W'N	4		
18	28.10	1430	bc	5	1	5	0	7	W'N	4		
18	28.10	1600	b	2	4	5	0	7	NW	4		
01	28.10	2000	b m	0	0	0	0	4	NW	2	44·7	07
[a] 01	29.10	0740	c	6	0	0	6	6	S'W	2	38·9	04
18	29.10	1340	o $r_0 r_0$	8	0	2	–	6	SW	4	52·8	12
01	29.10	1530	o rr m_0	8	0	2	–	5	SW	3	51·8	11
01	29.10	1715	o $r_0 r_0$	8	7	2	–	6	SW	4	50·0	10
01	29.10	1845	o	8	0	2	–	6	SW	4	51·6	11
01	30.10	0740	bc	4	0	5	0	6	WSW	2	52·0	11
18	30.10	1100	o	8	8	7	–	6	WSW	4	60·0	16
18	30.10	1450	bc	5	1	4	2	6	WSW	3		
01	30.10	1830	o id_0	8	–	–	–	6	W'S	3	57·5	14
01	30.10	2030	c	7	1	3	0	6	W'S	3	56·8	14
01	31.10	0715	o rr iR	8	0	2	–	4	NNE	3	49·7	10
01	31.10	1330	o rr	8	9	–	–	6	NE'N	4	47·5	09
01	31.10	1500	o rr q	8	9	–	–	6	N'E	6	45·8	08
01	31.10	2100	b	1	1	3	0	6	N'E	5	44·0	07

[a] Cirrus clouds moving from N.

Our own appears to be normal polar-air weather with a minor cold front. Next day, **28 October**, we stay in the ridge. Bar. pressure goes up to over 1020 mb., the wind remains in a north-westerly quarter, our cumulus clouds are small and flattened, while frontal-looking high clouds come to nothing. This ridge is doing a bit better than its predecessors to grow up into an anticyclone to block the way of the next Atlantic depression which now duly reaches Iceland with rain, snow, south-easterly gales and bar. falling as fast as 8 mb. per 3 hr. ahead of the front. Meanwhile the Genoa 'low' goes into the Balkans with thunderstorm followed by polar air blowing the Mistral (northerly gale) down the Rhone valley. Over France are scattered cumulus clouds all clearing at night, while one of the few high Alpine weather-reporting stations (Saentis, at about 8000 ft.) remains in cloud and snow below freezing-point.

29 October. Apart from a small separate centre of high pressure over south Norway this morning the ridge just fails to make it, and so the next warm front comes in—though not with much gusto here as the ridge still attains 1030 mb. over France which is near us.

The front here shows otherwise normal features, low and high-level-winds as you see reported at 0740, temperature rise from thirties to fifties Fahrenheit, frontal clouds and rain and finally C_L7 with poor visibility and rain then ceasing. West coasts in the warm sector have fog, with temperature 56° F. *Queen Mary* there steaming off westwards actually has temperature 60° F. with south-westerly wind (force 5) at midday. By evening, however, she is reporting overcast sky with Cb clouds, drizzle and *north* wind with force 1, which is followed by showers, north-westerly wind and temperature down to 53° F.

The cold front thereby located now slowly crosses our west coasts and then becomes almost parallel to the isobars (as the high pressure ridge extends west-south-west to east-north-east from Azores to France). This favours slight waves on the front. Quite apart from anything else the *Queen Mary*'s sudden increase not only of wind but of barometric tendency (from 0 to +3 mb. per 3 hr.) in the same air mass next morning is very suggestive that whilst in the cold air mass all the time she has passed underneath a slight upper-air wave crest soon after midnight (when at 49 N 23 W).

30 October. This wave thus east of 50 N 20 W in the morning duly comes up our south-west approaches, whose coastal weather reports show increasing rain and cyclonic circulation. Next morning, **31 October**, here it comes. 600 miles in 24 hr.—it has averaged 25 m.p.h., which again is nearly the warm-sector gradient wind speed. Strong cold winds now sweep round behind it to build the ridge up once more from the west. Those are our north-east to north winds to-day. *Queen Mary*, however, encounters another depression shown by a south-south-easterly gale and bar. falling 5 mb. per 3 hr. at longitude 40° W; so our 'high' is by no means secure yet.

7. ATLANTIC FRONTAL ANALYSIS

	Place no.	Date 1936	Time G.M.T.	Weather letters	Cloud				Vis. V.	Wind		Temp.	
					T.	L.	M.	H.		D.	F.	F.	C.
	01	1.11	0920	b m	0	0	0	0	4	W	2	41·0	05
	02	1.11	1200	b f	0	0	0	0	3	W	2		
	01	1.11	1320	c f	6	5	0	2	3	W	1	47·0	08
	01	1.11	1500	b m	0	0	0	0	4	W	2	45·9	08
	01	1.11	1630	b m	1	5	5	0	4	W	2	43·6	06
	01	1.11	2000	o rr	8	0	2	–	6	SW	3	45·0	07
	01	2.11	0730	o f	8	0	2	–	3	W	1	47·2	08
	18	2.11	1240	o m	8	3	2	–	4	NNE	1	52·1	11
[a]	01	2.11	1630	o m r_0r_0	8	3	–	–	4	N'E	3	48·0	09
	01	2.11	2150	b f	0	0	0	0	3	W	1	44·3	07
	01	3.11	0730	o	8	0	7	–	6	SW	3	45·1	07
	01	3.11	0750	o r_0r_0	8	0	2	–	6	SW	3	46·0	08
[b]	01	3.11	0820	o RR	8	0	2	–	5	SW	3	45·9	08
	18	3.11	1100	o rr	8	0	2	–	5	WSW	3	48·1	09
	18	3.11	1540	o f r_0r_0	8	0	2	–	3	WSW	2		
	01	3.11	2240	o f	8	0	2	–	2	Calm		48·2	09
	01	4.11	0800	c f	7	0	3	–	3	WNW	1	44·7	07
	01	4.11	1600	c f	7	0	3	–	3	Calm		47·2	08
	01	4.11	2200	o	8	0	3	7	7	SSW	2	45·4	07
	01	5.11	0750	c/r	7	0	7	6	6	SSW	3	49·5	10
	01	5.11	0820	o u	8	0	1	7	6	SSW	3	48·6	09
	18	5.11	1100	o r_0r_0 q_0	8	9	2	7	7	SW	4	50·8	10
	18	5.11	1230	c	7	9	7	8	7	SW	4		
	01	5.11	1720	c	7	5	0	0	6	SW'S	3	47·4	09
	01	5.11	2120	b m_0	0	0	0	0	5	S'W	1	43·5	06
	01	6.11	0740	c	7	0	5	0	6	SE'S	3	44·5	07
	01	6.11	0820	c	7	0	7	–	6	SE'S	2	45·1	07
	18	6.11	0930	bc	4	2	3	5	6	SSE	2		
	18	6.11	1340	c	7	2	0	6	6	S	3	50·2	10
	18	6.11	1400	c	7	3	5	6	6	S	3		
	18	6.11	1640	o	8	9	1	–	7	SSE	4		
	01	6.11	1730	o r_0r_0	8	9	2	–	7	S'E	4		
	01	6.11	1900	o rr q	8	7	–	–	6	S	5	46·8	08
	01	6.11	2130	bc	5	0	3	0	7	S	3	46·8	08
	01	6.11	2250	b	0	0	0	0	8	S	3		

[a] Gloomy sky full of London smoke.
[b] Rain becoming moderate.

1 November. Ships' weather reports indicate a preliminary occluding depression with warm front linked to our old cold front which, at first, is still over Spain but is soon off the map to the south-west as our polar air enters the trade wind zone of the tropics.

The occlusion causes our evening rain. 'Ridge' conditions ahead of it— light west-north-westerly drift of cold stable air (high Richardson number reducing turbulence) from industrial Midlands and London, with all the houses between them stoking domestic fires for a pleasant Sunday after-noon—produce the first really wintry London day, all quiet and foggy with smoke. It now needs not merely sunshine but unpolluted south-westerly wind to clear it, duly backing and freshening as the occlusion closely approaches. The picture looks more like that of 12 October than of any other day so far. A secondary cold front reaches Scotland whilst trailing in front of the next main depression (fourth of the series since 25 October when they started just as our series of charts in this book was ended) now going up to Iceland. The arctic whirl of the last one has just swept air down to Scotland from Greenland in contrast to the warmer maritime air over Britain behind the occlusion.

2 November observations show how warm it really is now after sub-sidence in the rebuilt high-pressure ridge. The secondary cold front, how-ever, arrives with rain and a sudden great concentration of London smoke as you see reported at 1630, duly followed by arctic air, cold and cloudless at night though still much polluted by smoke.

Iceland bar. meanwhile falls at the high rate of 6 mb. per 3 hr., and the occluding fronts of its new depression accordingly reach our west coasts.

3 November. The occlusion is bent back as on 15 October between P_m and arctic air. That is also how the secondary front over England has been, which now returns as a warm front giving us heavy rain as you see in the morning, followed not by rain ceasing as from a single front but continuing rain from the primary front to the south of it, whose warm sector has been truly ex-tropical air showing temperatures high in the fifties (F.). In-creasing rain and cyclonic wind circulation over south-western Britain, as on 30 October, show a slight wave developing. That is in a favourite place, the point where the warm and cold fronts at sea-level are joining to form the occlusion. We call it the POINT OF OCCLUSION.

4 November. Just as on 1 November, the ridge after the wave gives fog which is cleared by the south-south-west wind ahead of a new occlusion, which in turn has followed the last one round the Iceland low-pressure centre rather like that of 14 October. But now the 'high' growing up over south-east Europe slows everything down except north-eastwards. All fronts that have crossed us have piled up and slowly died out, lying south-west to north-east over Europe with widespread cloud and occasional rain.

The Iceland depression is showing a trough to the south, and a glance at the ships' reports of weather alone is suggestive. Two ships report heavy hail showers, and most ships report C_L9, implying abnormally widespread heavy convection such as is apt to develop a polar depression. Beyond it we note particularly the liner *Alaunia*'s weather reports. Steaming eastward at 53 N 37 W in the morning she reports cumulus clouds, temperature 41° F., a squally but slackening wind, and bar. at first falling, then steady, as though improving after being passed by a front or trough moving east in the usual way. As the ship itself is moving east, we infer the weather to be moving east even faster.

At 53 N 30 W in the evening the temperature has gone up to 45° F., the wind has backed round to the west, the cumulus clouds have flattened out into type 4, and thin altostratus (C_M1) has appeared. At midnight it is overcast with C_L7, occasional rain, and bar. falling 2·4 mb. per 3 hr., which is fairly fast.

5 November. At 53 N 25 W in the morning it drizzles in a south-south-west wind (force 5) while bar. has fallen about 4 mb. per 3 hr. Meanwhile the liner *Samaria* also steaming eastwards at 51 N 39 W at midnight likewise reports weather overcast with C_L7, occasional rain with west-south-west gale, temperature 52° F., and bar. falling 4 mb. per 3 hr. but changing by morning to fair weather (cumulus clouds) with temperature 42° F., north-west wind force 6, and bar. slowly rising. You see she has been just as clearly passed by a cold front as the *Alaunia* is being approached by a warm front; so a warm sector somewhere between 25 and 35 W must be coming after our polar depression (which with strengthening south-south-west winds and falling pressure in polar air is now recognised near our Western Isles). This new Atlantic affair, in fact, is as clear as the one we saw beginning on 16 October.

So much for the essential Atlantic frontal analysis first. Now from diagnosis to prognosis. Will things develop in the same way as they did before?

Let us see. The old occlusion ahead of the polar depression has reached us in London as you see by 1230 when rain has ceased. A wave on it, too, gives rain in south Norway, while another one starts over France where the older occlusion is suitably quasi-stationary; for with much high pressure built up over Russia no great advance of fronts into Europe is possible now.

6 November. After upper clouds just at first from the French frontal wave, we have polar air cumulus clouds until evening rain which is due to the occluded remnant of yesterday's well-marked Atlantic warm sector. At longitude 30 W which is about 1200 miles away, 40 hr. ago as you remember, it must have averaged 30 m.p.h., much the same as the speed of most other such fronts we have seen.

Our wind duly veers, but is still near southerly after the front. That is to say, the depression has not passed eastward but stays over Ireland. Moving east, you see, would mean invading the 'high', whilst moving west against all the westerly gales reported by the Atlantic ships would be out of the question. Except perhaps north-eastward, therefore, the 'low' is unlikely to move much at all, so we must expect quite a spell of south-westerly (mild P_m air) weather.

8. MINOR TROUGHS

Place no.	Date 1936	Time G.M.T.	Weather letters	Cloud				Vis. V.	Wind		Temp.	
				T.	L.	M.	H.		D.	F.	F.	C.
01	7.11	0745	o rr	8	0	2	–	7	SSW	3	44·2	07
01	7.11	0830	c	4	0	2	3	6	SW'S	4	44·0	07
01	7.11	0900	c pr	7	0	2	3	6	SW'S	3	45·0	07
01	7.11	1030	c pr	6	7	2	3	6	SW'S	5	46·1	08
01	7.11	1130	o r iR	7	0	2	3	6	SW'S	4	46·2	08
01	7.11	1215	c	7	0	2	2	6	SW'S	4		
01	7.11	1245	b	2	1	2	3	6	SSW	4		
01	7.11	1330	c pr_0	6	3	0	3	6	SW	5		
01	7.11	1915	b ic	1	0	0	1	6	SW	3	44·6	07
01	8.11	0845	b	1	0	0	1	7	WSW	3	46·6	08
01	8.11	1015	b	0	0	0	0	7	SW'W	5	47·3	09
02	8.11	1240	bc	4	2	0	0	8	SW'W	5		
01	8.11	1425	c pr	7	3	0	3	6	SW'W	5	47·7	09
01	8.11	1440	bc r	4	3	0	3	6	WSW	5	47·0	08
01	8.11	1530	c u pr	7	8	0	0	7	W'S	6	47·0	08
[a] 01	8.11	1720	b	1	5	0	0	6	WSW	2	42·9	06
01	9.11	0800	c/R	7	3	6	8	7	SW	6	46·3	08
01	9.11	1030	bc	3	1	0	8	7	WSW	6	50·0	10
20	9.11	1345	bc	4	2	0	8	8	WSW	4	52·0	11
01	9.11	1630	b	1	4	0	0	6	WSW	6	49·0	09

[a] Thunderstorm at 2230.

7 November. Fed by contrasting polar and arctic air, and/or by their convection, the 'low' goes on deepening over northern Ireland with gales blowing all round its south side. The general picture is slow to change, a general forecast for England simply being 'strong south-westerly winds with showers and otherwise good visibility'.

But the air's instability itself may as usual upset it by making such a lot of weather depend on such little things. In search of the little things, therefore, the forecaster seeks—and usually claims to find—minor fronts or troughs in the westerly air stream. His most useful weather reports then must be from Ireland, from ships or from aircraft over the Atlantic as far as possible up-wind (to south-west or west) of this country. 'Let's see the latest Irish reports!' may be the hourly cry in his office. In war-time, too, he wants south-west or western air meteorological reconnaissance sortie reports decoded and plotted as soon as received, just as he otherwise looks to ships. To-day's official analysis here, for example, traces a front as the back-bent occlusion arriving here after midnight; but any one of the day's rainy periods might likewise have been put down to minor fronts or troughs, though actually the weather after midday looks normal for P_m air without any fronts at all.

A front, you see, means a trough in the pressure pattern; but a trough does not always mean a front. Unlike a front, for instance, it by no means has to move with the air itself. Horizontal convergence into it may, how-ever, cause clouds and rain just like a real front, so that only its forecast *motion* differs. Although imitated locally (just a few miles at a time) by normal convection showers, any moving trough worth talking about can usually be detected by faster rises of pressure with squalls or veers of wind, as well as by extra cloudy or showery weather, all found to follow a line which smoothly advances round a depression centre, even though not with the speed of the wind.

8–10 November, as expected, are much the same. The average wind just veers gradually as the depression shifts slowly north-east.

9. 'TEXT-BOOK' DEPRESSION

	Place no.	Date 1936	Time G.M.T.	Weather letters	Cloud				Vis. V.	Wind		Temp.	
					T.	L.	M.	H.		D.	F.	F.	C.
	01	10.11	0750	c	7	0	7	–	6	W	3	43·9	07
	01	10.11	1200	c	7	2	0	9	7	W	4	48·0	09
	01	10.11	1400	bc/pr$_0$	3	3	6	3	6	W	5	46·0	08
	01	10.11	1800	b	0	0	0	0	6	WSW	4	42·0	06
	18	11.11	1050	o	8	0	0	7	6	SSW	2		
	18	11.11	1145	o	8	0	2	7	6	S'W	3		
	18	11.11	1230	o r$_0$r$_0$	8	7	2	7	6	S	4	47·2	08
	01	11.11	1340	o rr q$_0$	8	7	2	–	6	SSE	4	45·5	07
[a]	01	11.11	1545	o rr q	8	7	–	–	6	SE'S	5	45·6	08
	01	11.11	2220	c	6	–	–	–	8	SSW	5	51·6	11
	01	12.11	0730	o rr q	8	7	–	–	7	SSW	6	48·0	09
	18	12.11	1300	c d	7	9	–	–	6	SSW	5		
	18	12.11	1330	o RR	8	9	–	–	6	SSW	5		
	18	12.11	1440	bc	5	3	6	–	6	SSW	4		
	01	12.11	1715	o r$_0$r$_0$	8	7	2	–	6	W	4	50·0	10

[a] Temperature 52° F. by 2120.

10 November. While the old depression just sits raining off south-west Norway and all western Europe is covered with old P_m air, the first disturbance after four steady days has now appeared on the map far west. Steaming eastward at 53 N 36 W in the morning the liner *Empress of Britain* reports easterly wind and snow with bar. falling 3·2 mb. per 3 hr. *Bergensfjord*, too, steaming south-west at 47 N 51 W on the previous evening reported north-easterly wind with rain and C_M2 and bar. falling 5 mb. per 3 hr.

Those suggest a depression centred this morning not far south of 50 N, and not far west of 35–40 W. A ship steaming west at 49 N 29 W has south-easterly wind, temperature 55° F., and bar. falling 6·6 mb. per 3 hr., of which not more than about 3 mb. is accounted for by the ship's motion. *Bergensfjord* now in north-north-westerly wind has bar. *rising* 5·6 mb. per 3 hr., while other ships beyond 30 W have northerly winds which thus show a ridge coming after this new depression.

By evening the ships beyond only longitude 10 W report rain, which shows the approaching warm front. Contrast, for instance, *Empress of Britain* at 53 N (29 W) reporting strong *east* wind and temperature only 40° F. as against a ship at 48 N (26 W) reporting strong *west* wind and temperature up to 61° F., suggesting the warm front there lying more nearly east-west between their longitudes 53 and 48 N.

Next morning, **11 November,** the centre is clearly located by comparison of *Empress of Britain*'s cold northerly gale at 52 N 20 W (as also reported by liners *Laconia* and *Ascania* out to 53 N 35 W) with a warm westerly wind at 49 N 18 W and then cold south-easterly wind reported over our own south-western approaches with continuous rain.

Preceded, you notice, by normal warm frontal cloud types, this rain arrives here in London about 1230, the belt of it (200 miles in width) having come along fairly steadily at just over 25 m.p.h. Its rear edge accordingly gets here 8 hr. later, about 2030, followed of course by the warm front at ground-level bringing cloudy but fresh ex-tropical air with a veer of wind (from south-easterly to south-south-west) and quickly sending the temperature right up to 52° F. by 2120 G.M.T.

The centre of this depression passes London at 18 G.M.T. **12 November.** To estimate its kinetic energy is interesting. What mass of air moves, and how fast? Its weight is simply given by the bar. Between 976 and 1024 mb. the average pressure is 1000 mb., which in popular British units is about 2000 lb. per sq. ft. How many square feet then does this cyclone cover? From centre 976 to the 1024 mb. isobar is 800–900 miles. Actually this is

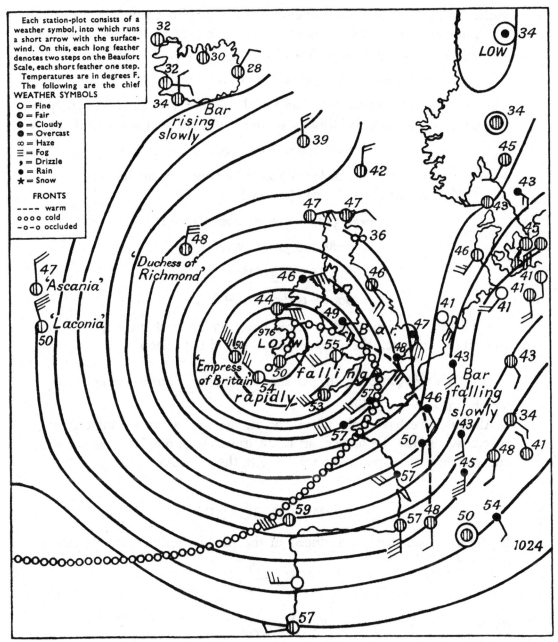

Fig. 45. A depression, 1800 G.M.T. 11 November 1936. This is a striking map of what you might well call a 'text-book' frontal depression crossing this country. On a larger scale than previous maps in this book, it is taken straight from the Air Ministry *Daily Weather Report* (International Section), showing not only isobars, fronts and wind directions but also wind force, weather and temperatures.

rather more than the main circulation area. Radius 600 miles out to the 1012 mb. isobar is more like it. So the area is $600^2\pi$ sq. miles. As a mile is 5280 ft., the area in sq. ft. must be

$$5280^2 \times 600^2\pi.$$

As the average pressure or weight of the air is nearly 2000 lb. per sq. ft. (a pound *weight* being 'g' times a pound *mass*), the total mass must be

$$5280^2 \times 600^2\pi \times 2000 \text{ lb.}$$

With *speed* then of the order of 100 ft. per sec., we finally get

$$\text{kinetic energy} = \tfrac{1}{2}(\text{mass}) \times (\text{speed})^2$$

$$= \tfrac{1}{2} \times 5280^2 \times 600^2\pi \times 2000 \times 100^2 \text{ ft. lb.}$$

$$= (5\cdot28^2 \times 10^6) \times (6^2\pi \times 10^4) \times 10^3 \times 10^4$$

$$= (5\cdot28 \times 6)^2 \pi \times 10^{6+4+3+4}$$

$$\fallingdotseq 1000\pi \times 10^{17} \fallingdotseq 3 \times 10^{20} \text{ ft. lb.}$$

If all this was generated in three days, which is about a quarter of a million ($\tfrac{1}{4} \times 10^6$) sec., the average power would have been

$$3 \times 10^{20}/\tfrac{1}{4} \times 10^6 \text{ ft.-lb. per sec.} = \frac{3 \times 10^{20}}{\tfrac{1}{4} \times 10^6 \times 550} \text{ horse-power,}$$

which is about two billion.

Although we have only taken 100 ft. per sec. as very rough average speed, we can use the same chart to illustrate measurement of geostrophic and gradient wind speed. So we come to the subject of WIND SCALES.

10. WIND SCALES

In the usual language, introduced in the Air Hydrodynamics chapter on Gradient Wind,

$$V_{gs} = \frac{\nabla p}{\rho L},$$

while

$$V_{gr} - V_{gs} = \frac{V_{gr}^2}{rL};$$

so that if

$$\frac{V_{gs}}{V_{gr}} = M,$$

then

$$1 - M = \frac{V_{gs}}{MrL}.$$

Now $\dfrac{1}{\nabla p}$ is simply the (perpendicular) *distance* or spacing between isobars drawn at unit intervals of pressure. A geostrophic wind scale for any particular values of ρ, L and map scale m may therefore be a straight line which is laid (along the radius of curvature) at right angles to the isobars and marked with velocities V_{gs} at distances $\dfrac{m}{\rho L V_{gs}}$ from one end.

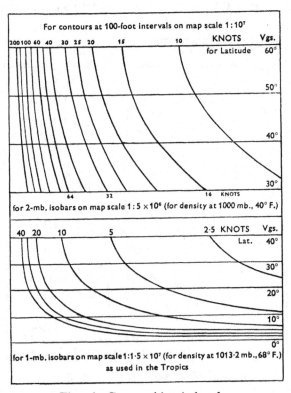

Fig. 46. Geostrophic wind scales.

A scale for all latitudes may then consist of parallel lines of this sort, say for 10° latitude intervals, the marked points for each V_{gs} being linked by a curve to facilitate interpolation.

The writer's method is first to make the scale in this way for latitude 90° and then to draw *perpendiculars* through its marked points, finally measuring (across the isobars as usual) not straight across these perpendiculars but at an angle ϕ to them, ϕ being the latitude. That is partly because in this direction the spaces between the perpendiculars are of course greater than on the 90° line in the ratio $1/\sin\phi$ in proportion to $1/L$ just as it should be.

In fact the latitude directions, along which to measure, may be marked by radius lines from one end of the 90° base-line. The other part of the reason for this design, quite apart from its neatness and adaptability to other things, is that as it is for a map having uniform map scale m ('distortion factor' $e=1$) you can adapt it to any standard map projection (meaning e not necessarily 1 but still a known function of latitude ϕ) by slightly curving the perpendiculars so as to change their distances from the 0° line in proportion to e, so that in effect they are graphs of the function e against

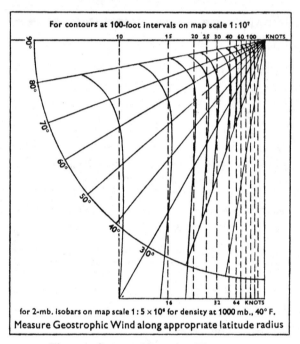

Fig. 47. Geostrophic scale. New type.

$\cos \phi$. For Meteorological Office maps, for instance, on conical projection with $e=1$ at standard parallels of latitude $\phi=30°$ or $60°$, it is as shown in Fig. 47.

Now for the *gradient* wind we have to take curvature ($1/r$) into account somehow. That is the curvature of the actual air-flow. It is only the same as the curvature of the isobars if the isobars are at rest with the wind blowing exactly along them. If nearly so, it can obviously be measured, though not by mere distance out from the low-pressure centre unless this happens to be the centre of curvature too, i.e. with circular isobars as on our chart for yesterday evening, 11 November 1936, where, for example,

$r=230$ miles at Belfast, and only 70 miles over parts of Wales, where $V_{gs}=50$ m.p.h.

The writer's general method of approximate curvature measurement is simply based on a well-known geometrical property of any arc of a circle. With the notation shown in Fig. 48,

$$d=\frac{h^2}{2r}.$$

If d is measured on the map while h and r are life-size, then

$$d=\frac{(mh)^2}{2mr}.$$

So if we fix what ('intercept') distance mh on the map shall be, fitting on to it any arc of a circle we find on the map (such as an isobar, approximately), then as in our diagram the short gap d will tell us the curvature radius r.

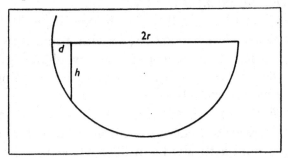

Fig. 48. Curvature measurement.

Now suppose instead of fixing this mh at right angles to the radius along which we measure the isobar spacing, we fix a distance mh at right angles to the $90°$ base-line. At this distance (to be conveniently chosen beforehand) our measuring scale is marked with a fixed line, labelled E, say. Or rather more precisely, as with the old 'perpendiculars', it is slightly curved to allow for the map projection. That is to say, if m is the nominal map scale, say 1 in a million (10^6) or whatever it is at the standard latitude, then the distance of E from the base-line is not mh but emh according to the latitude direction in which you measure.

That makes the intercept distance $\dfrac{emh}{\sin\phi}$, which is proportional to $\dfrac{emh}{L}$,

so that d is proportional to $\dfrac{\left(\dfrac{emh}{L}\right)^2}{2mr}$,

which thus measures $\dfrac{1}{L^2r}$.

Meanwhile the isobar spacing, say d', measures $\dfrac{1}{LV_{gs}}$, so that by suitable choice of h we can make d/d' measure V_{gs}/Lr.

Equivalently we can measure d on the V_{gs} scale, getting $V_{gs}=V_2$ say, a fictitious geostrophic wind, while the true $V_{gs}=V_1$. But at the beginning we saw that

$$1-M=\frac{V_{gs}}{MrL}$$

while now we see that

$$\frac{V_{gs}}{rL}=\frac{V_1}{V_2}.$$

Therefore

$$\frac{V_1}{V_2}=M\,(1-M),$$

whose graph against M (already plotted upon the measuring scale) will then tell us M and hence the gradient wind V_{gr}.

The scale shown in Fig. 49 is designed to give V_{gs} with isobars at 2 mb. intervals on map scale $m=1:5,000,000$; so with 2 mb. isobars on our map scale $1:20,000,000$ it would give $4V_{gs}$. V_{gs} would be given by 8 mb. isobars, such as 988–980 over Belfast on this occasion (evening of 11 November 1936), where we accordingly read $V_1=60$ knots.

Meanwhile designed to give V_2 on map scale $1:10,000,000$ it only gives $\frac{1}{2}V_2$ on our map here whose curvature distances d, if found on a one-in-ten-million map, would naturally be of lines whose real life-size curvature was twice as great as ours. On this larger-scale map, for instance, our 988–980 mb. lines would only look half as curved, and so would only have half the curvature distance d, giving wind V_2 just twice as strong as the one we had read. What we read is, therefore, only one-half of the true V_2. On our 984 mb. line through Belfast, for instance, this is also about -60 knots (being reckoned negative as usual with cyclonic curvature) so that $V_2=-120$ knots. Then $\dfrac{V_1}{V_2}=-\frac{1}{2}$.

If we neglect the cyclone's velocity c ($=15$ knots) in comparison with V_2 (thus taking the isobars' curvature to be nearly enough the true curvature of the air-flow), we read from the graph (middle curve, $c=0$) at 0.5 units down below centre-line (meaning that $M\left(1-M+\dfrac{c}{V_2}\cos\theta\right)$, which is V_1/V_2, is $-\frac{1}{2}$) near the right-hand side. The value of M is 1.36. As M means $\dfrac{V_{gs}}{V_{gr}}$, and in this case $V_{gs}=60$ knots, we deduce the gradient wind $V_{gr}=60/1.36$, about 44 knots.

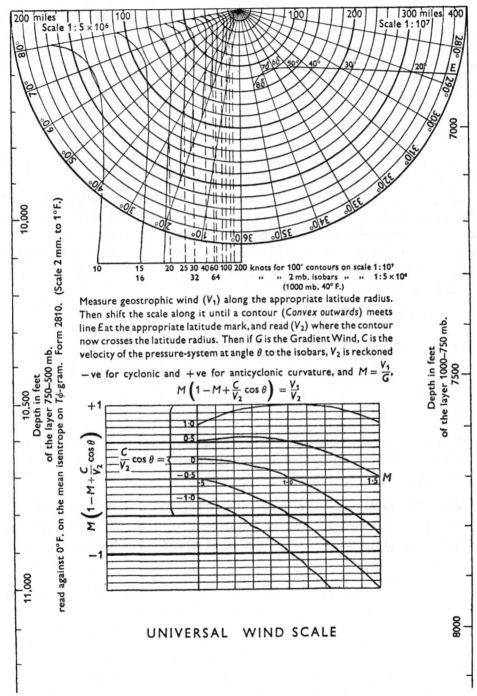

200 miles Scale 1 : 5 × 10⁶ ... 100 ... 100 ... 200 ... 300 miles | 400 Scale 1 : 10⁷

10 15 20 25 30 40 60 100 200 knots for 100' contours on scale 1 : 10⁷
16   32 64   "   " 2 mb. isobars "   " 1 : 5 × 10⁶
                        (1000 mb. 40° F.)

Measure geostrophic wind (V_1) along the appropriate latitude radius. Then shift the scale along it until a contour (*Convex outwards*) meets line E at the appropriate latitude mark, and read (V_2) where the contour now crosses the latitude radius. Then if G is the Gradient Wind, C is the velocity of the pressure-system at angle θ to the isobars, V_2 is reckoned —ve for cyclonic and +ve for anticyclonic curvature, and $M = \dfrac{V_1}{G}$,

$$M\left(1 - M + \frac{C}{V_2}\cos\theta\right) = \frac{V_1}{V_2}$$

10,000

10,500

Depth in feet
of the layer 750–500 mb. Form 2810. (Scale 2 mm. to 1° F.)

read against 0° F. on the mean isentrope on Tϕ-gram.

11,000

7000

Depth in feet
of the layer 1000–750 mb.

7500

8000

UNIVERSAL WIND SCALE

Fig. 49. Universal wind scale.

That is very different from V_{gs}. As it is so much nearer to the true wind, you see how important it is in this case to try to allow for the curvature.

Over Wales at midnight only one-third as far out from the centre as Belfast was, with curvature therefore about three times as great (V_2 being now only one-third of -120, while V_1 itself is about 40 knots), you see

$$\frac{V_1}{V_2} = -\frac{40}{40} = -1.$$

So we have to go off the right-hand edge of the graph and estimate M to be just over $1\cdot5$; V_{gr} then is only about 25 knots.

Many ways of thus allowing for curvature in deducing the gradient wind have from time to time been devised; but to the writer's knowledge this is one of the easiest, if not the only one having everything on the one measuring scale. This original scale was engraved by hand upon perspex for use in the office, and then reproduced by photography. All combined in the one set of lines are:

Geostrophic scale for use with isobars.

Geostrophic scale for use with contours.

Geostrophic scale for use with standard map-projection but adaptable to others.

Scales of miles, 1 in 5 million, 1 in 10 million.

Polar graph (radius distances, arcs and angles marked in degrees for graphical vector addition or subtraction, such as for estimation of thermal winds).

The scales down the sides, of depth of the 1000–750 and 750–500 mb. layers of the atmosphere, will be explained, together with contours and other topics, in the chapter on upper wind analysis and forecasting.

IX
FRONTS AND ANTICYCLONES

1. DOUBLE FRONTS

	Place no.	Date 1936	Time G.M.T.	Weather letters	Cloud				Vis. V.	Wind		Temp.	
					T.	L.	M.	H.		D.	F.	F.	C.
	01	13.11	0730	o d q	8	5	–	–	6	N′W	5	47·0	08
	18	13.11	1100	o	8	5	–	–	6	NNW	3	48·2	09
	18	13.11	1315	c	7	3	6	–	6	NNW	3		
	01	13.11	1600	c m	7	3	–	–	4	NW′N	2	46·5	08
	01	13.11	2130	c f	7	–	–	–	3	WNW	2	43·8	07
[a]	01	14.11	0730	c	7	0	7	–	6	S	3	43·3	08
	01	14.11	0750	o pr	8	0	7	–	6	S	3		
	01	14.11	1100	bc/iR	2	0	4	4	6	WSW	3	47·7	09
	01	14.11	1330	bc	3	1	7	4	6	WSW	4	49·2	10
	01	14.11	2100	c	7	0	0	6	6	WSW	3	42·4	06
	01	15.11	0815	bc/b	5	0	0	6	7	WSW	3	40·7	05
	01	15.11	0900	c	7	0	7	6	6	WSW	4	43·2	06
[b]	01	15.11	1330	o	8	7	7	7	6	SW	5	50·0	10
	01	15.11	1500	o r_0r_0	8	7	2	–	6	WSW	4	49·5	10
	01	15.11	1620	o	8	7	7	7	6	SW′W	5	49·0	09
	01	15.11	2140	o d	8	5	–	–	7	WSW	6	50·5	10
	01	16.11	0740	o	8	8	2	–	6	W′S	3	50·6	10
	18	16.11	1100	o	8	3	1	–	6	W	3	53·2	12
	18	16.11	1230	c	7	1	7	8	6	W	3		
	18	16.11	1330	c d	7	0	1	–	5	W	2		
	01	16.11	1730	o/rr m	8	0	2	–	4	S	1	51·5	11
	01	16.11	2140	o d_0>rr	8	0	2	–	6	SW	2	53·0	12
	01	17.11	0745	o	8	5	–	–	7	SW	4	53·3	12
	18	17.11	1100	o r_0r_0	8	7	–	–	7	SW	4	55·8	13
	18	17.11	1445	o r_0/rr q	8	5	–	–	7	WSW	5		
	18	17.11	1540	c	7	5	–	–	7	SW′W	3		
	18	17.11	1640	o rr	8	7	–	–	7	SW′W	4		
	01	17.11	1940	o RR	8	7	–	–	6	WSW	4	52·7	11
[c]	01	17.11	2010	o m R>rr q	8	7	–	–	4	W′N	4	52·7	11
	01	17.11	2230	o	8	–	–	–	6	NW	3	47·6	09

[a] Medium clouds moving from west.
[b] Medium clouds moving from WNW.
[c] Wind force rising to 6 while temperature falls to 47° F. by 2030.

13 November. The 'textbook' depression centre having passed us by yesterday evening, 12 November, we duly find ourselves in a northerly stream of air which must have curved over Europe and right round Britain,

143

thoroughly soaked and cooled down by the rain. A ridge, of course, comes in from the west, though unfortunately for us who are fond of the more settled weather of anticyclones, this is one of the cold and transient sort. The high-pressure centre in arctic air east of Iceland is only a 'hillock' with not more than one closed isobar drawn. Linked across Britain with a ridge to the main subtropical high-pressure belt, it moves quickly eastward over a whole degree of longitude every 2 hr. without growing up, and so by **14 November** it has passed us, and southerlies blow here once more to bring us a front. The trailing front of the last depression, whilst entering Europe three days ago, was forming a second wave depression on the Atlantic, which became just as intense as the last one, but went up to Iceland from which it is now almost fully occluded. We see its occlusion this morning followed by mild P_m air with fair-weather cumulus clouds for a few hours until a secondary cold front sweeping after it gives us another cold night.

By **15 November** both fronts are waving. The secondary one gets its blow in first to bring the main low-pressure centre right out of Iceland and quickly south-east as its warm front approaches with normal features until arrival at ground-level by 2140 to-night as you see.

On **16 November** the cold front has cleared us far enough to allow cumulus clouds to appear in the wedge of cold air underneath the frontal altostratus clouds, but soon it is coming back as a warm front again with normal features including rain, drizzle, wind veer, visibility change and temperature rise by 2140. The other front farther south is catching it up and forcing it north as it develops a wave depression itself. Returning to England in the morning, for instance, the *Aquitania* with south-west wind force 7 and temperature up to 59° F. is evidently in an ex-tropical warm sector.

17 November. We ourselves are obviously in the warm sector until about 1430, when cold front goes through. But our system of fronts, remember, has been not single but double. We could call them primary and secondary. Such double structures, well marked if respectively polar and arctic fronts, though otherwise less so, are quite common. Official pictures may 'pool' them, showing just one front half-way between them. That makes a nice simple picture, but hardly enough for forecasts in detail. Double warm fronts in summer, for instance, are apt to vary so much in relative effect that you hardly know where they are, or which is which. Double cold fronts are generally more distinct, though the secondary one may be rightly or wrongly traced as a back-bent occlusion.

To-day's, anyway, should be no surprise. We see it at 2010 with heavy rain, wind veer, squall and sudden temperature drop, after which we notice the wind go right round to the north-west and then next morning (18 November) a northerly gale.

In this connection a common fault of beginners in frontal analysis is the location of a front upon different charts by different features which may not all be changing in the same place at once. Clouds and weather, for instance, arising aloft, do not always change where the *ground*-level wind or temperature do. The squall line is not always the edge of the rain belt. For an airfield controller requiring warning of squalls or sudden changes of wind across his runway you must remember that the effective front is the squall line, which should therefore be located largely by observations of wind and thereby forecast for him as a squall line alone. But for someone who wants to know when it will rain—or stop raining—the effective front instead is the edge of the rain-belt, which should be located and forecast as such. Comparison with its position on a previous chart, say 60 miles back at 3 hr. ago, gives its speed, 20 m.p.h. For a place say 200 miles ahead its arrival is therefore not due until 10 hr. later. Comparison instead with the position of the squall line on the previous chart, say a further 60 miles back, total 120 at 3 hr. ago, gives fictitious speed 40. So its forecast time of arriva 200 miles ahead would be 5 hr. out, a serious error. The different effective fronts, in short, must not be mixed up.

2. ANTICYCLONIC WEATHER

18 November. The 'high', which on the last occasion looked so promising but then gave way to the oncoming cyclone, now comes in, so to speak, by the back door. Having really come in, it will not be pushed out. Its cold centre, linked by a ridge to the old subtropical 'high' centred near the Azores, now appears intensifying over Norway, from which our north-easterlies start to prevail. A cyclone deepens off Iceland to 956 mb. but fails to advance. Its south-westerly gales are forced to stream for thousands of miles straight from the Azores to the arctic ice. Over London our old frontal rain goes on most of the day as the depression (having moved east from England) slows down and stops still quite near us before being sent back to the south-west by our new prevailing air stream.

This goes on till midday **19 November**, when subsidence under the rising pressure must have so weakened the front that, as it also goes away to the south-west, it stops raining upon us. Over Germany the precipitation goes on much longer, turning to snow. The 'high' is now centred 1037 mb. over the Gulf of Bothnia, forming a source of polar (P_C) air that starts its journey towards us at about 0° C. under the influence of the snow that is falling from remnants of ex-maritime air high aloft on the occluded front, where persistence of very low clouds all day shows almost complete

Place no.	Date 1936	Time G.M.T.	Weather letters	Cloud T.	L.	M.	H.	Vis. V.	Wind D.	F.	Temp. F.	C.
01	18.11	0740	o	8	5	–	–	6	N	6	44·0	07
18	18.11	1100	o d/rr	8	5	–	–	6	N	5	45·8	08
01	18.11	1330	o r_0r_0	8	9	–	–	6	NNE	5	43·1	06
01	18.11	1800	o	8	9	–	–	6	NNE	6	44·0	07
01	18.11	2215	o	8	5	–	–	7	NNE	4	44·8	07
01	19.11	0740	o m_0	8	7	–	–	5	NE'N	3	43·9	07
18	19.11	1000	o r_0r_0	8	7	–	–	6	NE'N	4		
18	19.11	1100	o d	8	7	–	–	5	NE	2	45·0	07
01	19.11	1700	o	8	7	–	–	6	NE'N	3	43·1	06
01	20.11	0750	b m_0	2	0	0	4	5	NNE	3	38·9	04
18	20.11	1100	c	7	5	–	–	6	NE'E	4	43·9	07
01	20.11	1840	o z	8	0	1	–	4	NE'E	3	41·0	05
01	21.11	0100	c	7	0	3	–	6	NE	2		
01	21.11	0745	b m	1	0	0	1	4	NE	3	36·5	03
18	21.11	1100	c m_0	7	1	0	0	5	NE'E	4	42·7	06
01	21.11	1320	b	1	1	0	0	6	E	3	43·7	07
01	21.11	2100	b f	0	0	0	0	2	E	1	37·7	03
01	22.11	0920	o F	8	5	–	–	0	E	2	38·9	04
19	22.11	1220	b f	0	0	0	0	2	ESE	3		
01	22.11	1230	b F	0	0	0	0	1	SE	1		
01	22.11	1400	b f	0	0	0	0	3	SE	3	43·0	06
01	22.11	1450	b m_0	0	0	0	0	5	SE	3	42·9	06
01	22.11	1600	b f	0	0	0	0	3	SE	2	40·8	05
01	22.11	2030	o F	8	5	–	–	1	E	2	36·8	03
01	22.11	2040	o f	8	5	–	–	2	E	3	35·8	02
01	22.11	2210	o f	8	5	–	–	2	E	1	36·2	02
01	23.11	0750	o F	8	5	–	–	0	NE	2	36·3	02
18	23.11	1100	o f	8	5	–	–	2	NNE	1	34·8	02
18	23.11	1250	o f	8	5	–	–	2	N	1	37·0	03
01	23.11	1700	o f	8	5	–	–	3	W	1	38·0	03
01	23.11	2220	o f	8	5	–	–	3	S	1	38·1	03
01	24.11	0800	o m	8	5	–	–	4	SSE	1	37·1	03
18	24.11	1100	o m	8	5	–	–	4	SSE	1	36·5	03
18	24.11	1610	o m	8	5	–	–	4	SE	1		
01	24.11	1740	o f	8	5	–	–	3	SE	1	36·6	03
01	24.11	2240	o f	8	5	–	–	3	Calm		37·1	03
01	25.11	0820	o f	8	5	–	–	3	N	2	36·2	02
18	25.11	1145	o f	8	5	–	–	3	N	1	34·8	02
01	25.11	1540	o f r_0r_0	8	5	–	–	3	N	2	37·0	03
01	25.11	1845	o f r_0r_0	8	5	–	–	3	N	2	38·0	03
01	25.11	2115	o f r_0r_0	8	5	–	–	2	N	2	39·0	04

saturation, the dry temperature being little higher than wet temperature which is about freezing-point.

On **20 November** the main high-pressure centre, shifting with customary ease along the ridge towards the old Azores 'high', reaches England. That is nothing to get excited about, of course, the ridge having already settled the weather into an anticyclonic type. Being November, it means a cold spell with no rain but much fog. No rain to speak of, anyway. Every household barometer, duly pointing to nearly 31 in. of mercury pressure, says 'Very Dry'.

Away in Iceland, on the other hand, household barometers now pointing to only 29 in., with the words 'Rain' or 'Stormy' only too true, must imply that with pressure so much lower on their left hand than on their right hand, all the Atlantic south-westerlies are now pouring into the Arctic regions to give them even warmer weather than ours in latitude 50° N. In fact there are times when it even snows at sea-level at 30° N. (on the continent of Asia) whilst quite mild at 60° N. on the west coasts of Europe.

Any friction between the arctic south-westerlies and our north-easterlies, as at the edge of a river, should tend to set up clockwise (anticyclonic) eddies which might then keep up our 'high'. Of the first of these arctic systems just now the occlusion (3000 miles long from Azores to Spitzbergen) comes down as near us as north England, where it dies out in the high-pressure centre. All Britain's weather is fine by midnight.

On **21 November**, apart from some short-lived cumulus clouds, the characteristic initially cloudless weather of an anticyclone is seen to begin, followed of course by night radiation fog. The sunshine is feeble now, and the night is long. Once formed, by **22 November**, the fog cannot be warmed or dried away except just a bit in the afternoon, when it thins out to visibility No. 5 (2–4 km.). Note, by the way, how it alters with the force of the wind.

In the evening we see an equally characteristic feature of a winter anticyclone. That is a stratus or stratocumulus cloud layer, formed underneath an inversion probably from convection off the North Sea up to 1–2 km. When smoky from London or any big city it darkens the daytime sky with 'anticyclonic gloom'. In contrast to the westerly gales over northern Europe the rest of the continental air is now likewise stagnant with fog and frost, here overcast, there clear. All old fronts die out with subsidence, so that neither rain nor snow nor medium clouds nor cumulus clouds are left anywhere in the anticyclone, even when it has so declined that sea-level pressure is almost uniform.

A second depression has meanwhile crossed Iceland over to Spitzbergen in the Arctic, trailing a front from the White Sea all the way to the Orkney Isles and thence in a wave west of Britain to the Azores, behind which, of course, a new anticyclone develops.

23, 24 November show no change here in London. The overcast sky now cuts off even the little heat of the sun, so the fog is more settled than ever. All Europe, in fact, lying south of the front found initially over the White Sea (behind which the cyclone sequence is stopped by a flood of arctic air from the north-west) has weather like ours. The wave to the west of us hardly develops at all; nor, with opposing 'highs' now to east and west of it, can it move further. What happens is that the 'highs' join up on its north side so as to send it back off the map to the south-west. If you refer to the map for 8 October you will see the same sort of picture.

On **25 November** our own weather shows just one new feature. It rains. The change of wind from south-east to northerly since yesterday might also be noted. The official picture, however, remains featureless with neither any marked fronts nor isobars anywhere over Britain.

Something subtle is happening. However lightly, it does not rain without reason. Drizzle it might, but not rain. Post-mortem examination reminds us first that the wave to the west of Britain was frontal, and that the front, like the one before it, was over 3000 miles long from Azores to beyond the White Sea. The occluding wave in it may have been swept away south-westward all right, but can such a front itself be so simply disposed of?

We have suggested the answer Yes. We say it dies out in the 'high'. But quite apart from whether or not this is true, we have to remember that for simplicity's sake a front may be dropped from the official picture before it is dead. Rain-dot symbols are sprinkled over to-day's map of Europe without any sign of depressions, troughs or isobars at all, the anticyclone having become a vast region of almost uniform pressure.

3. MINOR FRONTS AND VIRTUAL TEMPERATURES

Place no.	Date 1936	Time G.M.T.	Weather letters	Cloud				Vis. V.	Wind		Temp.	
				T.	L.	M.	H.		D.	F.	F.	C.
01	26.11	0815	o f d	8	–	–	–	3	NNE	1	40·2	05
18	26.11	1100	c m	7	0	6	–	4	ENE	1	42·0	06
01	26.11	1700	c m	7	3	6	–	4	E	1	41·0	05
01	26.11	2200	c f	7	5	–	–	2	E	1	38·6	04
01	27.11	0745	o f	8	5	–	–	2	N	1	37·5	03
18	27.11	1100	o f	8	5	–	–	3	N	1	39·0	04
01	27.11	2100	o f	8	5	–	–	3	Calm		39·9	04
01	28.11	0750	b m	1	5	0	0	4	WNW	1	36·1	02
18	28.11	1100	b f	1	5	0	0	3	WNW	1	34·0	01
14	28.11	1430	c m	6	2	7	0	4	NW	2		
01	28.11	1900	c m	7	0	7	–	4	NW	1	40·7	05
01	28.11	2130	c m	7	5	–	–	4	NW	2	40·8	05
01	28.11	2300	c m_0	6	5	0	0	5	NW	3	39·0	04
01	29.11	0920	o f	8	0	7	–	3	SW'S	3	36·6	03
19	29.11	1220	o rr	8	0	2	–	4	SW'W	2		
01	29.11	1315	o $r_0 r_0$	8	0	2	–	5	SW	4	40·8	05
01	29.11	1350	o d_0	8	7	2	–	5	SW'W	4	40·7	05
01	29.11	1600	o d_0	8	7	–	–	4	SW'W	3	42·4	06
01	29.11	1720	o/rr	8	7	–	–	6	WSW	4	43·9	07
01	29.11	1910	o RR	8	7	–	–	6	W	3	44·6	07
01	29.11	2100	o m_0	8	5	–	–	5	W	3	46·0	08
01	29.11	2230	o	8	5	–	–	6	W	3	47·0	08
01	30.11	0730	o	8	7	7	–	6	W	3	48·6	09
18	30.11	0900	c	7	0	7	9	6	W	4		
18	30.11	1100	c	7	0	7	9	6	W'S	3	49·5	10
18	30.11	1230	c	7	1	7	–	6	W'S	4		
18	30.11	1400	o	8	8	–	–	6	WSW	3		
18	30.11	1445	bc	4	2	1	9	6	W	3		
01	30.11	1630	c	7	8	0	9	6	NW'W	3	50·0	10
[a] 01	30.11	1920	b	1	4	0	0	6	NW'W	3	46·5	08

[a] 22 G.M.T. wind WNW.

'All old fronts', it was said in the previous section, 'die out...so that neither rain nor snow nor medium clouds nor cumulus clouds are left anywhere in the anticyclone.' Does a front really die out in a 'high' like this? 'The activity of a front', it was said in an earlier chapter, 'is least at a high pressure centre, where the air is almost calm. High pressure also causes general outflow over the ground...the upper air flows in downwards and so counteracts up-flow on a front, thus tending to clear the bad weather on the spot.'

Fair enough. But what next? 'Low pressure makes things worse. Minor cold fronts, for instance, are apt to cause heavy showers round a depression whilst innocuous round an anticyclone.' *No front on its way to us, however innocuous-looking, is therefore to be neglected.* Even though it need not keep on cluttering up the official picture, still we must trace it lest the pressure should fall. Even now it may not be so very innocuous either. At least it remains as a marked change of moisture or smoke concentration, so that here, where a little of these now means so much to the weather, it may easily make all the difference. A minor front always does something. In fact it is often the only way to explain or anticipate certain remarkable changes of weather from what you would expect at first sight of the picture. If taken then as their explanation it can be located by them. On 7 November, for instance, 'the general picture is slow to change, a general forecast for England simply being "Strong south-westerly winds with showers and otherwise good visibility". But the air's instability itself may as usual upset it by making such a lot of weather depend on such little things. In search of the little things, therefore, the forecaster seeks... minor fronts or troughs in the westerly air-stream. His most useful weather reports then must be from Ireland, from ships or from aircraft over the Atlantic as far as possible upwind (to south-west or west) of this country. "Let's see the latest Irish reports!" may be the hourly cry in his office....'

'Although imitated locally (just a few miles at a time) by normal convection showers, any moving trough worth talking about can usually be detected by faster rises of pressure with squalls or veers of wind, as well as by extra cloudy or showery weather, all found to follow a line which smoothly advances round a depression centre.' *That includes minor frontal as well as non-frontal troughs.*

Be careful, however. You are warned against hastily marking a front wherever stations reporting showers are found to lie more or less in a line. There might be a line of hills there which would account for it just as well. Or the only available stations might happen to lie in a line anyway. Don't be misled, and remember never to mix up different effective fronts.

In air which is not from the south-west or west but like to-day's from the Continent (north-east or east), on the other hand, our most useful reports are from far inland stations where it is drier. A minor front in fine weather might be marked just by an altocumulus cloud layer or by sudden haze thickening. In fair weather it might be marked by a bank of Cb among Cu clouds. In cloudy weather it might be marked by a short deterioration with low stratus clouds and drizzle. In any weather, in short, it can probably be detected by an otherwise unexpected development of precipitation and/or clouds. Rain, for instance, apart from fronts, is normally only due to deep damp air rising on mountains or in convection. Convection requires instability; so in the absence both of instability and of mountains we may infer any rain to be frontal. Always we may, and usually we must. The most sensitive indicators are clouds; big Cu, for instance, growing anywhere far inland at night (apart from high hills) being almost certainly frontal. All these may only be signs of a minor front; but these very signs themselves—sudden banks of cloud or deteriorations in visibility—all matter a lot to those who fly, to whom any failure of our anticipation is a serious error that may even be fatal. That is why we should never neglect a minor front. Remember, too, that it never dies. It may be asleep, but the slightest disturbance can wake it.

Like most things in nature, no two fronts are found exactly the same. All they have in common are the basic features we already know, with e. cold air underneath e. warmer, or, rather more precisely, not warmer but what we call *virtually* warmer. What really counts, as you must have noticed, is the air's *density* (ρ). Did *temperature* (wet, dry, potential, equivalent or any other sort) appear in our air's hydrodynamics, in any wind formula? No; only ρ.

Yet surely we need to relate them, ρ and temperature T. The simple gas law tells us the pressure

$$p = R\rho T,$$

R being constant. Moist air is lighter than dry air, water vapour being only about five-eighths as dense as dry air under the same conditions. Its density, say ρ_w, is $\frac{5}{8}\rho$ for the same p and T, so you see that its corresponding constant

$$R_w = \tfrac{8}{5}R$$

so that if the air is $100x\%$ water vapour, whose density therefore is $x\rho$, then the total pressure must be

$$R\rho T + R_w x\rho T$$
$$= R\rho T\,(1 + \tfrac{8}{5}x)$$

at which, of course, the same amount $(1+x)$ of equally dense but perfectly dry air would then have a temperature equal to

$$\frac{R\rho T\,(1 + \tfrac{8}{5}x)}{R\rho\,(1 + x)}$$
$$= T\left(1 + \frac{\tfrac{3}{5}x}{1 + x}\right)$$
$$\simeq T\,(1 + \tfrac{3}{5}x)$$

which is the VIRTUAL TEMPERATURE.

That, then, is the temperature which would give the air the same density if it were perfectly dry, so that now our hydrodynamics need never explicitly mention the moisture.

26 November shows the changes brought by our minor front from yesterday. Clouds lift and break slightly, the fog thins out, the light wind veers round to the east, and the temperature rises higher than ever. Passing north of Britain the high pressure centre moves east now to Russia. From the Azores to the Arctic a third occlusion of this series has now appeared, with yet another high-pressure surge on the opposite side. Next day, **27 November**, this surge successfully pushes forward the front against our slowly retreating European 'high'. Failing actual precipitation the front is dropped out of the official picture, whereat it promptly turns up and clears us by the evening of **28 November**. Behind it the fourth occluding Atlantic system of this series sweeps forward past Iceland. On **29 November** the development of the Atlantic 'high' gives this one a different track from the others before it. Moving round this dominant 'high' instead of a European one, the fronts come along from the west-north-west like those illustrated on 11 October.

First we see a warm front with normal features and plenty of rain. Then, just as we saw illustrated before on 12 October, the next low-pressure trough comes even more freely. Iceland bar. falls very rapidly ahead and then rises behind it. We see the trough arrive here in London next day, **30 November**, about 14 G.M.T. There is no rain, but just a briefly overcast sky and then a wind veer. On the October occasion the front hung back just off Land's End, which was so close to us as to make the weather stay cloudy. This time it trails through the high-pressure centre nearly a thousand miles away to the west-south-west, so the weather clears up. The next Atlantic system does not happen to have followed closely enough to have pulled our front back yet. Our last November evening is, therefore, almost unclouded in polar air, fresh and clear after our week of anticyclonic gloom.

As a historical note, our final remark is an observation of wind direction which was made at 22 G.M.T. on the drift of the smoke from the burning Crystal Palace.

4. ANTICYCLONES AND THEIR MOTION

The picture now resembles not only that of 13 October but also (though less closely) those of 18 and 27 October, to which you can refer. Our 18 October system was the fifth of its series, while that of 27 October was only the second in the following series. Ours of to-day at the end of November is again the fifth of a series but only the second to come to Britain.

In each previous case an anticyclone afterwards did its best to spread north-eastwards to slow down or stop these Atlantic fronts coming our way at all. Will this happen again? It happened that in the two previous cases the anticyclone advanced in different ways, which made all the difference to us; yet to the picture as a whole the difference was only subtle. 'On the map as a whole', as we said before, 'you could not go far wrong, but to any one place it might easily make all the difference....'

You see how hard it is even to forecast the *type* of weather two or three days ahead with a picture like this. Anticyclones are indeed subtle. Just as a local change of only a few feet in ground-level, insignificant in a deep valley, nevertheless makes all the difference to a stream meandering over a plain, so will a small change in pressure, insignificant in a deep depression, make all the difference to the direction of air-flow in the more nearly uniform pressure of anticyclones. Quietly yet quickly, for instance, it can shift the high pressure centre a thousand miles overnight, and thereby deflect all approaching depressions and fronts. Cyclones may be all wind and fury and yet fail to dominate the situation. It is the placid *anticyclone* that largely controls them. Studies of anticyclones like this have suggested long ago that their important changes come from high up. As this is partly true of cyclones too, we come back now to an opening question of this book. *How is the atmosphere built?*

Sufficiently high ascent always shows that at some level normally 5–10 miles above the earth the temperature ceases to fall much—if at all—with height. Above this level, up to a much greater height, it is nearly constant. This level also appears as the topmost limit of all marked haze and moisture. No more visible on a fine day than the ionosphere without which you would have no long-distance radio, this level is nevertheless a definite 'ceiling'. Our weather is all on the 'ground floor' beneath it. That is the TROPOSPHERE. The ceiling is known as the TROPOPAUSE, and the 'first floor' is the STRATOSPHERE.

As temperature only falls with height as far as the tropopause, we may infer that (given the air mass) the higher the base of the stratosphere is, the colder it is, and therefore the colder and denser must the whole stratosphere be, and so *the more pressure will it exert*. That is assuming, of course, that the air is piled up and has not yet had time to spill away out of the higher stratosphere. We find it very high over the tropics but lower over the poles, high over anticyclones ('highs') whilst low over cyclones ('lows'). Tropical convection, horizontal convergence or anything else that piles up air in the ground floor, necessarily tends to raise the ceiling and so to cool the first floor. Of ex-tropical air, for instance, the first floor should therefore be cold. The tropical ground floor is of course warm, so that their resultant tropical sea-level pressure is about average. But ex-tropical over ex-polar air in higher latitudes means cold first floor over cold ground floor, so that resultant sea-level pressure is high. That gives us an anti-cyclone. Remember the permanent subtropical high-pressure belt, where fine dry weather is so everlasting that most of the land underneath it is desert.

In still higher latitudes, at the polar front, ex-tropical and ex-polar air meet not only in their respectively warm and cold ground floors, but also in their respectively cold and warm first floors, over respectively high and low tropopause ceiling. Keeping low to the left as usual, upper winds therefore blow from a westerly quarter. Gradient, too, being steep, these upper westerly winds are very strong in a belt which narrows in places into a stream whose speed may exceed 200 knots. Such JET STREAMS are important in air navigation. Obviously, for instance, they favour fast west-east transatlantic flights, can seriously upset the operation of bombers, or may even carry volcanic dust round the world. Look out for them high on the polar front within 400–700 miles of the sea-level front, or even within 200–400 miles, often marked by the speed of frontal cirrus clouds over the sky. Upper-air isobars, high-level winds, stratosphere isotherms, tropopause contours and the troposphere's polar front itself are then all nearly parallel. As ex-tropical air comes north-eastwards round the west side of any subtropical anticyclone (like to-day's or the one in October somewhere near the Azores), all these lines just ahead of it veer in direction, becoming roughly north-west to south-east. That implies high-level pressure highest to the south-west. In so far as the sea-level centre of highest pressure is no further north than this, it is not far enough north to cut us right off from the advancing ex-tropical winds and fronts. It just steers them round into our west to north-west current. But if the veer is right round to a north-south direction, then high-level pressure is highest to west, so that even though sea-level high pressure centre is no further north than our own latitude it may now cut us off from Atlantic air after all; and if the veer should go farther round to north-east to south-west we are almost assured of a spell of continental weather instead.

With north-westerly or even with northerly high-level winds, however, as seen in advance of a warm front from the Atlantic, the pattern with highest pressure to south-west or west may not be an anticyclone but only a ridge, the main anticyclone being farther south, right out of the way. Then we are fully exposed to an average westerly air-stream. Above the warm front the winds all back from north of west to south of west, the ridge being followed by a trough. Ridge and trough, of course, form a wave; and a wave is something that easily moves. Air masses at a warm front converge horizontally and so diverge vertically. In so far as they make up for this by converging vertically higher up, the warm air stream will diverge horizontally and so acquire anticyclonic spin. Over the cold front the opposite process goes on, and so the wave deepens. Not that it is really as simple as this; but you see how the causes may come down from above, as well as up from below. The high-level pattern is something like that of Fig. 21, for which an eastward motion was explained. (Or if seen the same way up on the map in southern latitudes it would be a westward motion.) In the end (if above an unstable polar-front wave) it develops a cyclonic centre. Then as the 'cyclostrophic' wind component diverges out of the rear and converges into the front part of any moving cyclone, pressure tends to be raised ahead and lowered behind the centre, thus retarding the centre.

Around the moving cyclone, you see, the cyclostrophic component is opposite to the geostrophic wind, whilst proportional to the curvature of the air-flow. So on the starboard side, where the curvature of the air-flow is less than that of the isobars, the wind must differ from that which has been deduced from the isobars alone, by a component which blows the *same* way as the geostrophic, whereas on the port side it differs by a component which *opposes* V_{gs}. Both of these components thus blow across the instantaneous positions of the isobars, out of the rear and into the front half of the cyclone. Being slower, however, than the isobars' own advance, their motion *relatively to the moving isobars* must be backwards from the front to the rear.

By the same argument round a moving anticyclone, where the cyclostrophic wind blows the same way as V_{gs}, we infer that upon the port side, where curvature of air-flow is less than that of the isobars, the wind must differ from that which the isobars give, by a component *against* V_{gs}, whereas on the starboard side it differs by a component *with* V_{gs}, so that on both sides the difference is a component across the instantaneous isobars from front to rear.

So we see horizontal convergence of air behind an anticyclone but ahead of a cyclone, helping in each case to pile up the air and so raise the pressure, reducing the barometric tendency and thus slowing down the whole system.

The system's bodily movement (say with velocity c) contributes to the barometric tendency what we have called the convective or advective term, the rest being the deepening term. Obviously the convective term is just c times the pressure-gradient component along the line of motion. With isobars inclined to the motion at angle θ, for instance, it is simply

$$c\nabla p \sin \theta.$$

Picture the isobars as contours of a hilly surface. Then $\nabla p \sin \theta$ gives the projection of unit area of this surface on to any plane at right angles to the motion. In a beam of light along the line of motion, for instance, the intensity of illumination of the surface would vary as $\nabla p \sin \theta$, which would therefore be greatest upon the steepest hillsides facing the light. Here then are the ISALLOBARIC (barometric tendency) centres, into or out of which the isallobaric wind components blow. So we see how a pressure-system's bodily motion alone creates isallobaric winds, whose horizontal convergence and divergence tend to reduce the maxima and increase the minima of barometric tendency. If, in short, the isallobars are pictured as contours of hilly country instead, the country becomes flattened out.

In circular isobars equally spaced, for example, curvature k and ∇p being uniform, the isallobars of the convective term must be lines of equal θ, namely radii of the circles. So the isallobaric gradient, along which the isallobaric winds blow, being at right angles to the isallobars, must be along the circles, along the isobars themselves. Distance in this direction being radius $(1/k)$ times the angle turned through, whilst the barometric tendency is

$$c\nabla p \sin \theta,$$

the gradient of barometric tendency convective term alone must have the mathematical formula

$$\frac{\mathrm{d}\,(c\nabla p \sin \theta)}{\mathrm{d}\,(\theta/k)}$$

$$= kc\nabla p \cos \theta.$$

156

The isallobaric wind, then, proportional to

$$kc\nabla\mathrm{p}\cos\theta,$$

having horizontal divergence

$$\frac{\mathrm{d}\,(kc\nabla\mathrm{p}\cos\theta)}{\mathrm{d}(\theta/k)}$$

$$=-k^2c\nabla\mathrm{p}\sin\theta$$

representing *convergence* at a rate exactly proportional to the barometric tendency itself, should have the effect of retarding the isobars' motion bodily without alteration of their shape.

So we have begun to analyse how isallobaric as well as cyclostrophic wind components may affect the motion of pressure-patterns. If we could also allow for friction (horizontal as well as vertical), we should be even nearer the truth. Jet streams, for instance, may set up great horizontal eddies, developing cyclones on one side and anticyclones on the other. But though we might hope thus, by hydrodynamical theory alone, at last to predict all the air-flow, and so to predict the synoptic picture in detail a long time ahead, it is so complicated to deal with the changing patterns at different levels all at once, and the air is itself so often unstable, that beyond a certain stage in our progress the hope must be vain. You may like to know, however, that weather prediction by hydrodynamical calculation has long been the subject of expert research along various lines. All the lines enter forests of problems blocking our way to the hoped-for peak of perfection; but some have already gone far, and one of the newest lines follows the idea of using electronic 'brains'

5. FURTHER WAVES AND FRONTS

Place no.	Date 1936	Time G.M.T.	Weather letters	Cloud T.	L.	M.	H.	Vis. V.	Wind D.	F.	Temp. F.	C.
01	1.12	0830	bc	4	0	3	0	6	W	3	40·5	05
18	1.12	1100	c	7	8	—	—	6	W	3	44·4	07
01	1.12	2015	c	7	0	3	0	6	NW'W	4	41·0	05
01	2.12	0800	o rr	8	7	—	—	6	WSW	3	44·4	07
18	2.12	1100	o $r_0 r_0$ q	8	9	—	—	6	W	5	52·0	11
18	2.12	1300	c	7	8	—	—	6	WNW	4		
01	2.12	1730	o	8	8	—	—	6	WNW	3	50·8	10
01	2.12	2230	o	8	8	—	—	7	WNW	4	51·0	11
01	3.12	0730	o	8	0	3	—	6	W	3	48·7	09
18	3.12	1100	b	2	1	3	0	6	W	4	51·8	11
18	3.12	1230	c	7	8	3	0	6	W	3		
01	3.12	2110	o	8	5	—	—	7	WSW	5	50·4	10
01	4.12	0750	b	0	0	0	0	7	W	4	45·1	07
18	4.12	1100	b	1	1	0	0	7	WNW	5	47·8	09
18	4.12	1540	b	0	0	0	0	6	W'N	3		
01	4.12	2000	b	0	0	0	0	6	SW	3	39·0	04
01	4.12	2130	c m_0	6	0	5	—	5	SW	2	39·7	04
01	5.12	0810	o rr	8	9	2	—	7	WSW	4	43·0	06
18	5.12	1100	c d_0	7	9	7	—	7	WSW	3	42·5	06
18	5.12	1145	c	7	3	4	0	6	WSW	4		
01	5.12	1315	bc	5	3	0	3	6	W	3	43·9	07
01	5.12	1335	c u ps_0	7	3	—	—	5	W	4		
01	5.12	1430	bc ps_0	4	3	6	3	6	W	3	40·0	04
01	5.12	1445	b	1	3	6	3	6	W	3	39·9	04
01	5.12	2210	bc	3	4	6	0	7	W	3	35·0	02
01	6.12	0900	o rs/s	8	9	—	—	6	W'N	3	37·0	03
01	6.12	1000	o ss	8	7	—	—	4	WNW	3	36·0	02
01	6.12	1020	o rr	8	9	—	—	4	WNW	3	35·5	02
01	6.12	1240	c	7	8	6	2	6	WNW	3	38·0	03
01	6.12	1420	c	6	2	6	0	6	NNW	5	38·3	03
01	6.12	1720	b	0	0	0	0	6	NNW	3	35·0	02
01	6.12	2240	b	0	0	0	0	6	WNW	3	34·0	01
01	7.12	0730	b	0	0	0	0	6	W	3	31·5	00
18	7.12	1300	b	0	0	0	0	6	NW	2	34·0	01
01	7.12	1645	b m	0	0	0	0	4	N	1	31·0	−01
01	7.12	1840	b f	0	0	0	0	2	WSW	1	30·0	−01
01	7.12	2120	b f	0	0	0	0	3	SW	1	28·0	−02
01	7.12	2210	b f	0	0	0	0	3	S	1	27·5	−03
01	8.12	0730	o $r_0 r_0$ f	8	7	—	—	3	SW	1	33·9	01
18	8.12	1100	o m	8	7	—	—	4	SW	4	38·9	04
18	8.12	1230	o m rr	8	7	—	—	4	SW	4		
01	8.12	1715	o f rr	8	7	—	—	3	W	1	42·8	06
01	8.12	1900	o f	8	7	—	—	3	NW	3	41·4	05
01	8.12	1945	c f	5	7	—	—	3	NNW	4	41·0	05
01	8.12	2145	o m	8	7	—	—	4	N	1	40·5	05
01	9.12	0745	b f	0	0	0	0	2	W	1	33·1	01

[a] at the row: 01 | 4.12 | 2130

[a] Medium clouds moving from W.

Place no.	Date 1936	Time G.M.T.	Weather letters	Cloud				Vis. V.	Wind		Temp.	
				T.	L.	M.	H.		D.	F.	F.	C.
18	9.12	1100	o F	8	5	–	–	1	Calm		35·1	02
01	9.12	1920	c f	8	5	–	–	2	NW	2	38·5	04
01	10.12	0750	o m_0	8	5	–	–	5	NE	2	37·6	03
18	10.12	1100	o m_0/d_0	8	9	–	–	5	ENE	2	35·5	02
18	10.12	1340	c m_0	7	8	–	–	5	E	2		
01	10.12	1900	c m/ps_0	7	8	–	–	4	NNW	2	37·4	03
01	11.12	0750	bc m	3	0	0	9	4	S	1	31·0	−01
18	11.12	1100	c m_0	7	0	0	6	5	S	2	38·5	04
18	11.12	1300	c m_0	7	0	0	8	5	S	2		
01	11.12	1750	b m_0	1	0	0	1	5	S	2	34·0	01
01	11.12	2150	c	7	–	–	–	6	SSE	2	34·2	01
01	12.12	0800	o m	8	5	–	–	4	SE	2	35·8	02
18	12.12	1100	o m	8	7	–	–	4	SSE	2	40·6	05
01	12.12	1340	o m d_0	8	0	2	–	4	S	2	40·7	05
01	12.12	1740	o rr m_0	8	7	–	–	5	S	1	40·6	05
01	12.12	2240	o r_0r_0	8	7	–	–	5	NW	2	38·6	04
01	13.12	0910	b m	0	0	0	0	4	SSW	3	32·4	00
01	13.12	1315	b	0	0	0	0	6	SSW	3	40·5	05
01	13.12	1500	b z_0	1	1	3	0	5	SSW	3	40·9	05
01	13.12	1640	c pr	7	3	6	3	6	SW	3	40·5	05
01	13.12	1930	bc q	5	2	0	0	7	SW'S	4	41·5	05
01	13.12	2030	c q	6	2	0	0	7	SSW	5	42·9	06
01	13.12	2215	c q	7	4	1	–	9	SSW	6	43·9	07
01	13.12	2250	o r_0r_0 q	8	5	2	–	9	SSW	6	44·2	07
01	14.12	0800	o DD Q	8	5	2	–	7	SSW	7	48·0	09
18	14.12	1100	o rr q	8	7	–	–	7	SSW	6	49·5	10
18	14.12	1250	o rr q	8	7	–	–	7	S'W	5		
01	14.12	1630	o rr q	8	7	–	–	5	S'W	8	47·3	09
01	14.12	1850	o rr	8	7	–	–	6	S'W	7	47·6	09

1 December. Atlantic warm front appears and is steered round the 'high' into our west to north-west current.

2 December. We see it arrive with normal features and big rise of temperature by 11 G.M.T. It forms part of a wave which deepens little. That is to say its cold front does not sweep south-eastwards after the warm front but remains quasi-stationary (east-south-east to west-north-west) trailing straight into the next depression's warm front. The mere fact that it is stationary, however, enables us to distinguish it from the advancing first half of our slight frontal wave, and so to locate the otherwise indistinct wave trough (or wave crest as seen on a map with north at the top, though really the trough both of pressure and of the warmer air-stream over the front). The change, you see, is of *weather type*, from the *rainy* type at first in the wedge of cold air under the part of the front that is 'warm' (i.e. advancing slowly *against* the cold air side), to *fair* weather type (perhaps still cloudy but no longer with rain) in the same cold air wedge but under the part of the front that is quasi-stationary. There is no change of air mass, you see; yet the trough line may be located at once where rain-dot symbols leave off upon the synoptic chart. To-day it is marked thus by rain dots seen first in the morning over northern Scotland and then in the evening over the Low Countries. That is 480 miles in 12 hr., showing an average speed of 40 m.p.h. the same as the warm air gradient wind, as we should expect. Yet the isobars show hardly any trough from the main Scandinavian 'low'. The 'high' remains to the south-west. We in London, you notice, are far enough south to have entered the warm air mass. As the cold front is not coming, we can expect to stay warm for some time.

This situation is characteristic of any wave in an initially warm type of front, as opposed to the more familiar case of a wave in an initially cold type of front.

3 December. The next warm front now advances in earnest with a depression deepening about 5 mb. per 3 hr. and approaching the Shetlands from west. Deepening and occlusion are both rapid (illustrating a general rule that the one depends on the other), the cold front overtaking the warm front by about 50 miles every hour, and so clearing us by next morning, **4 December**, when fine weather with lower temperatures clearly show polar air.

The centre has moved at 40 m.p.h. parallel as usual to the isobars of the warm sector, whose gradient wind has been 40 m.p.h. in the south whilst 70 in the north of Great Britain. The centre now slows down to 30, but gales round it go on unabated with southerlies sending the fronts right off to the Baltic while north-westerlies sweep right down from the Greenland Sea where temperatures average zero Fahrenheit, near -20° C.

5 December. In this air stream a depression (presumably polar) appears and rains on us early, followed by cold polar air with the first snow showers. Its front appears back-bent, presaging real arctic air to follow.

6 December. Here it comes. London roofs are white before their first real winter's day ends cloudless and frosty. A ridge is following from the newly intensified 'high'.

7 December. The sun is so low in the winter sky that although quite unclouded (convection being almost impossible on such cold ground with such high pressure) it cannot warm much above freezing-point. What we gain in daylight with such clear skies we more than lose in heat. In contrast to possible maximum temperature of about 30° C. here in London by May, 20° C. in March, 15° C. in February and 5–10° C. on fine days in January, sunshine alone cannot make shade temperature much above 0° C. in December.

Slowly but steadily comes the next Atlantic warm front to reach Wales at midnight with coastal temperature rising to 10° C. in contrast to London's several degrees of frost. Not many warm fronts in this country are so well marked at ground-level, and of these there are few which meanwhile appear so inactive. The secret appears to be a small wave moving south past Land's End, leaving part of the front temporarily stationary and therefore inactive like the one on 2 December. Next day, **8 December**, the rest of the front comes on and obviously arrives in the afternoon, though promptly followed by cold front with wind-veer to northerly, showing a new ridge to follow with anticyclonic weather on **9 December**.

How will this anticyclone develop? The picture shows main 'high' centre (1045 mb.) to south-west with 'low' to north, from which the fronts across this country are trailing out westward as on 30 November, when we were introduced to the difficulties of forecasting anticyclones and their motion. The Atlantic 'high' is connected to one over Russia by a long ridge in which, for example, Swiss Alpine stations are sending reports of cloudless skies as we should expect, and from which we can picture scenes of great winter beauty. Never forget that weather reports leave much to your imagination! You can imagine the Russian winter anticyclone from a single report at Sverdlovsk in the Urals. Bar. 1040 mb., sky cloudless with minimum temperature −20° C.

This is the kind of picture in which we might see northerly upper winds here finally veering north-east. 'If the veer', remember, 'is right round to a north-south direction, then...even though sea-level high-pressure centre is no further north than our own latitude, it may now cut us off from Atlantic air after all; and if the veer should go further round north-east to south-west, we are almost assured of a spell of continental weather instead.'

Compare 22 August, when we could see the upper wind but not the synoptic picture. For to-day we have the description of the picture, but no report of the upper wind. Drawing upon experience of occasions with both upper-wind and synoptic reports we can just reverse what we said on 22 August. 'A new warm front lying nearly north-south' suggests to-day 'that the wind veers a lot with height...from a light wind coming from west of north to a strong upper wind that comes from north.'

Sure enough, as on 23–24 August, next day's weather (**10 December**) is anticyclonic, though with slight precipitation as you see, suggesting as

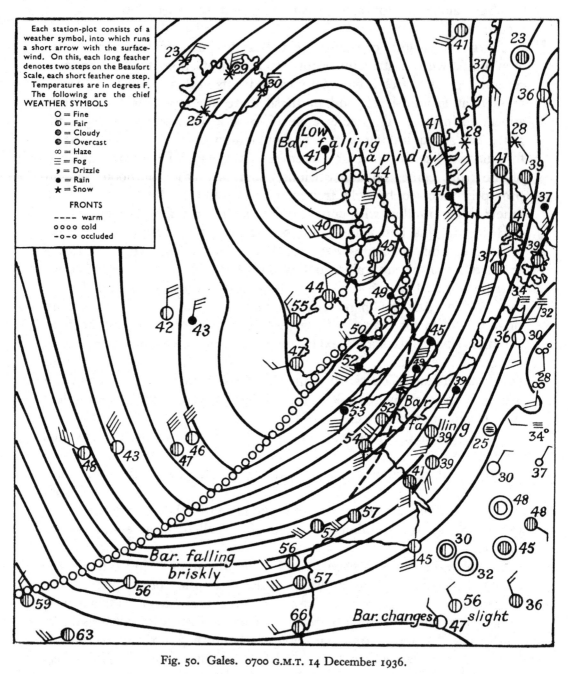

Each station-plot consists of a
weather symbol, into which runs
a short arrow with the surface-
wind. On this, each long feather
denotes two steps on the Beaufort
Scale, each short feather one step.
Temperatures are in degrees F.
The following are the chief
WEATHER SYMBOLS

O = Fine
⦶ = Fair
⦵ = Cloudy
⦿ = Overcast
∞ = Haze
☰ = Fog
𝄈 = Drizzle
● = Rain
✳ = Snow

FRONTS

---- warm
oooo cold
-o-o occluded

LOW
Bar falling
41   *rapidly*

Bar falling

Bar falling briskly

Bar changes slight

Fig. 50. Gales. 0700 G.M.T. 14 December 1936.

usual a minor front. It is, in fact, our last occlusion coming back now with the easterly wind. Meanwhile the next occluding front is trailing as on 22 November 'all the way from the White Sea to the Orkney Isles and thence in a wave west of Britain to the Azores, behind which of course a new anticyclone develops'.

This one, however, instead of taking a week to reach us, advances on **11 December** like that of 22 October, where you may remember that 'with the high pressure centre so near us we see only a few low clouds'.

On **12 December** again as on 22 October, 'the charts show a big cold front...a slow-moving front that must be carefully watched for the slightest sign of a wave. To-day's, however, develops no wave, so it duly arrives...', this time in the evening, as rain is reported and the wind veers round from south to north-west.

On **13–14 December,** as on 24 October, 'the front having passed', we have 'an ordinary day of cumulus clouds in the P_m air until evening when upper clouds start to suggest the approach of the first warm front of a new series. This is next observed in vigorous action on the following morning'. Then as on 3 December, 'the next warm front now advances in earnest with a depression deepening about 5 mb. per 3 hr. and approaching from west'. Deepening is again rapid, but occlusion is not, as a slight wave develops to hold up the main cold front. That will probably leave us in the warm air stream all next day, with cloudy weather and gales.

X

UPPER WINDS AND FORECASTING

1. ISENTROPIC AND ISOBARIC ANALYSIS

In this striking picture of gales coming over this country on 14 December, 1936, a remarkable feature is the absence of upper clouds at only 300 miles or about 500 km. from the sea-level warm front. There must be more in it than meets the eye. How shall we find out?

Let us picture just what is going on. A mass of e. warm wet air is curling over our own hitherto e. cold dry air. What happens? 'The air's most conservative property', you remember, 'is wet potential temperature (w.p.t.).' 'Next best', you then remember, 'are dry potential temperature (d.p.t.) and moisture content (m.c.), being unaltered by any adiabatic ascent or descent as long as the air is not saturated.' That is to say that these particular properties of moving air remain unchanged under more general changing conditions than any others. But among our first principles in Chapter I, 'any change in any property of a particular moving particle or element of a fluid is called a TOTAL change...and then any change of weather can be analysed into two parts':

    I. 'Total' change

    II. 'Advection', or BODILY TRANSPORT of properties by the flow.

In so far, then, as 'total' changes of d.p.t. and m.c. are zero, *weather changes depend on their transport alone*. So we have only to know their distribution and motion. That is the idea of isentropic analysis: just as air-mass analysis deals with the distribution and motion of w.p.t.

Now remember what potential temperature or d.p.t. is. It was defined as the temperature *reduced to* or *taken up at* some standard pressure. Reducing it to some standard pressure, like reduction to sea-level, is a device employed to allow for height. What is the effect of height? Temperature falls with height at a rate known as lapse rate, and we are particularly concerned with air that is 'self-supporting', exchanging no heat with the air around it. As ascent increases its potential energy it must reduce its heat energy and so cool down. 'This way of cooling is given a name. We call it adiabatic. Descending air likewise is warmed.' One property is constant. That is the air's ENTROPY.

What is entropy? Beginners in meteorology, when interviewed by their seniors, are sometimes purposely asked this single but shattering technical question, just to see what they make of it.

Two nice ways of putting it were once made by Sir Arthur Eddington. He suggested that you take a new pack of cards, as received from the makers. After a few minutes' shuffling you not only lose all trace of the original order, its *organisation*, but you will not get it back. Not by mere further shuffling, anyhow. You have done something you cannot undo. You have introduced RANDOMNESS into the pack.

There is just the ghost of a chance of your happening on the original order again, but we are going to deal with things of this sort in which so many are shuffled together that such a chance can be disregarded. We are reminded of the idea of monkeys strumming on typewriters, happening on exactly the right arrangement of letters, etc., to reproduce quite by chance all the books in the British Museum. It is not impossible but too improbable, though the probability may quite definitely be stated in figures. Entropy is in effect a short way of stating it.

Quoting from earlier in this book in a different connection, 'the Uncertainty Principle only a generation ago for the first time showed the very idea of any perfectly certain prediction whatever (however long taken for granted) to be an illusion'. That will at least suggest to you how universal is entropy's importance.

Eddington next invites us to recall Humpty Dumpty, who sat on a wall, where by virtue of his height above ground he had some potential energy.

Humpty Dumpty had a great fall. Was the energy lost? Not at all. It went into the shivering of every atom of Humpty's shell, reappearing as heat.

Yet:

> All the King's horses
> And all the King's men
> Couldn't put Humpty Dumpty together again.

So *something* must have been lost. What could it have been?

His falling body was made of molecules, atoms, all moving together like an army on the march, as Eddington recalls:

> The noble Duke of York
> He had ten thousand men;
> He marched them down a hill and then
> He marched them up again....

No, that is not quite the right quotation—but you see it is what Humpty Dumpty's atoms ought to have done. Had they kept all their organisation like the marching army, moving down to the bottom and then bouncing up again to the top, they would have restored Humpty Dumpty to the top of the wall. As they failed, we can now see what they must have lost. It was ORGANISATION.

When asked 'What is energy?' what have you always been taught to reply? 'Energy', you are taught to say, 'is capacity for doing work.' That is clearly not true for Humpty Dumpty. His energy was not lost, yet it could not restore him to the top of the wall—not even if harnessed to a steam-engine or other machine for the purpose. Capacity for really doing work is surely better described by the energy's organisation. As the loss of capacity is clearly measurable, so must the loss of organisation be. *Entropy is a measure of energy's disorganisation.*

Disorganisation, or rather organisation, is something you may or may not recognise in a set of things. In a sense it is therefore relative. A wall-paper pattern inside a house might at first sight appear as much at random as rough-casting on the outside wall, and yet by carefully comparing strip with strip you might detect that the pattern repeats after all, showing thereby less complete disorganisation than appeared at first sight. Disorganisation, in fact, is relative to your recognised standard, rather as we now recognise concord in what early musicians regarded as discord. That is how, as Eddington summed it up, entropy is an *idea* more of the nature of harmony or of beauty than a mere *thing* of the nature of energy. More, however, than beauty, or even than musical harmony which is already largely a matter of wave numbers, the idea of entropy is so largely a matter of figures related to the scientific characters of energy, temperature, etc., that we accept it as another respectable character in our science, if not indeed one of the very foundations.

The particular science in which it concerns us here, of course, is meteorology. After defining 'adiabatic' in our first principles we said that 'a simple law then relates the temperature with pressure and so with height. The rate of change of temperature with height is the so-called adiabatic lapse rate.' The entropy of the air is equivalent to, or rather represented by, the so-called POTENTIAL TEMPERATURE obtained by reducing actual temperature to 1000 mb. level at the (dry) adiabatic lapse rate.

So now we do not talk about just plain hot or cold air, but speak of *potentially* hot or cold air, height having been allowed for at this standard rate. Any remaining potential temperature variation with height must mean actual lapse rate different from this standard. Super-adiabatic lapse rate, for instance, means air potentially colder above than below, and therefore liable to overturn. That is convection. The entropy or potential temperature of the rising air itself becomes uniform. Most of the atmosphere, however, is potentially warmer above than below, being stable if dry (unsaturated). We shall not complicate the issue yet with saturation, clouds or 'e.' temperatures.

A line or surface through places of equal potential temperature, being of uniform entropy, is called an *isentrope*. Most isentropes therefore lie flat (or with just gentle slopes, hills and valleys) above one another, with the potentially warmest on top. Warming the upper air, as for example by subsidence, lowers them, crowding them into 'inversions' or stable layers. Cooling it raises them, spacing them higher above one another as the lapse rate approaches adiabatic, until, when potential temperature becomes uniform all the way up, we have all the different isentropes reduced to just one. With dry convection over land temperature 30° C., for instance, the 30° C. isentrope is ideally no longer a surface but the whole space up to the inversion or stable layer into which the other isentropes have been crowded aloft.

So you can picture them, rising dome-like, for instance, over the 'cushion' of cold air formed inland at night, or rather over a great land mass in winter so that a land breeze or winter monsoon slides down their slopes to the sea. By day or in summer, as it gets potentially warmer inland than at sea, the isentropes' hill turns into a valley, down whose slopes there then blows a sea breeze or summer monsoon. In fact, even at night in summer when local slight cooling inland near the coast may raise a few isentropes into a cliff or rather low hill shape, the monsoon will go on blowing inland, even though slowed down. *By virtue of the air keeping constant d.p.t. as it goes along* (having total change zero, as we have already seen) *and so keeping along the isentropes*, it may then have to slide up hill above level ground just as if the ground itself were rising. Hence bad weather. As we have to predict the regions and movements of bad weather, why not do it in terms of the hills and valleys of isentropes? Obviously it changes with them. It moves as *they* move. If they moved horizontally with the air, of course, other things being equal, the air would have little or no occasion to slide up or down their slopes, and so there should not be much weather. But generally the gradient wind or air's horizontal motion is not quite the same as the isentropes'.

Although dry air may keep to the slopes of its isentropes, it is not necessarily down the directions of these slopes that it finds itself pushed, nor along their contours that it finally flows. Down whose slopes then *is* the air pushed? We know that in the first place it is pushed from high to low pressure. Here again, of course, we must take the height into account. Pressure, like temperature, falls with height. Unlike temperature, in fact, it must. So we must remember to take the pressure at one height at a time. Its isobars then are horizontal.

Now just as HORIZONTAL surfaces mark out height, while ISENTROPIC surfaces, one above another, mark out potential temperature, so ISOBARIC surfaces may be defined to mark out pressure. There is one for 1000 mb. normally just above sea-level, one for 700 mb. about 3 km. up, and so on.

Our horizontal isobars, then, are simply the lines where *isobaric* surfaces cross a *horizontal* surface.

The lines where *horizontal* surfaces cross an *isobaric* surface would logically be called ISOBARIC CONTOURS.

Both are intersections of horizontal and isobaric surfaces. So they are equivalent. Just as the gradient wind is along the horizontal isobars, so is it along the isobaric contours. That is the actual air-flow, *as seen in plan* on a horizontal map.

The weather, however, depends on the actual air-flow up or down the slopes of the isentropes, i.e. relatively not to the isobaric contours but to the ISENTROPIC CONTOURS. So it remains for us to compare isobaric with *isentropic* surfaces. First consider where they intersect. Just as the isobaric and *horizontal* surfaces meet in *horizontal* isobars, so the isobaric and *isentropic* surfaces meet in *isentropic isobars*. Of pressure, temperature and potential temperature, any two will determine the third; so isentropic isobars are also isotherms and isosteres. As they meet the *isentropic* surfaces' contours where they meet the *isobaric* surfaces' contours, which, in turn, are the air's true (or gradient wind) flow lines as seen in plan, we deduce that along any true flow line on an isentrope, as we go up the standard height interval from contour to contour we go down the standard temperature interval from isotherm to isotherm, so that our STREAM FUNCTION constant along any stream line should be the sum of a multiple of the temperature and a multiple of the height. The temperature variation with height being adiabatic (10° C. per km.), this stream function may be taken as

$$\psi = T + 10H,$$

where temperature T is in ° C. while height H is in km.; or rather more precisely

$$\psi = T + 10\Phi$$

where Φ, the GEO-POTENTIAL, just g times the height, is in 'dynamic km.' which exceed ordinary km. in the ratio 1000/981, g being 981 basic units.

Note the dimensions:

$$1 \text{ dynamic cm.} = \frac{1000}{981} \times 1 \text{ cm.} \times 981 \text{ cm./sec.}^2$$

$$= 1000 \text{ cm.}^2/\text{sec.}^2$$

$$= 1000 \text{ basic units of dimension } \mathbf{L^2/T^2},$$

which is energy ($\mathbf{ML^2/T^2}$) per unit mass ($\mathbf{M}$), which is what we mean by potential.

Knowing T and Φ all over the place, you see, we can work out ψ's, plot them and so draw ψ lines to see how they cross the isentropes' contours up hill or down. The air itself flowing along the ψ lines will accordingly slide up or down hill unless either there are no crossings or else the whole isentrope moves horizontally with the air.

So much for isentropes, which are of d.p.t. What about m.c., the air's equally conservative property?

In our picture for 14 December 1936, with warm air isentropes like a deep valley, m.c. might be marked on each isentrope by lines for values, say 0·1, 0·2, 0·3...% of the air. *These will mark the air itself*, inasmuch as they move with the air until carried so far up the slope of each isentrope as to become *saturation* m.c. lines, beyond which of course the air no longer keeps constant d.p.t., but only constant w.p.t. That means it would keep to w.p.t. surfaces (for which, in default of any recognised name, the abbreviation 'e.' *surfaces* is here suggested, as they mark where the air is potentially e. warm, e. cold, etc.), but would thereby rise through the d.p.t. (isentropic) surfaces. Ideally with uniform w.p.t. all the way up, such a rising air mass would automatically keep constant potential e. temperature whilst rising from isentrope to isentrope. If, on the other hand, all its moisture then fell out as rain, any descent would be dry (unsaturated) and therefore isentropic again.

Such is isentropic analysis. For weather forecasting it may help us little or no more than frontal or air-mass analysis, to which you will see its equivalence. But quite apart from its simpler connection with pressure contours and temperature patterns for gradient wind flow lines, it long ago showed a promising feature in the form of its *moisture* patterns revealing great horizontal (or rather isentropic) eddies, such as may be set up by jet streams and might not otherwise have been detected or analysed.

Beyond this introduction, however, we need not pursue isentropic analysis here but will turn instead to the more familiar *isobaric* method.

Our construction of ψ lines through the intersections of T lines and Φ lines is known as GRIDDING, because from these three sets of lines you see that any two sets will form a network or grid that determines the others. 'Absolute' contours, for instance, of height (say H_A) of the 700 mb. surface above sea-level, are likewise constructed by gridding those of the 1000 mb. surface height (say H_O) with the 'relative' contours that mark the thickness (say H_R) of the layer of air in between; for obviously

$$H_A = H_O + H_R,$$

just as $$\psi = T + 10\Phi.$$

As the absolute 700 mb. surface contours show geostrophic (or gradient) wind at their level, so do the absolute 1000 mb. contours show V_{gs} at theirs, and so therefore do the relative contours or THICKNESS LINES SHOW THE THERMAL WIND. The thickness, you see, is simply the depth of air that exerts 300 mb. pressure, representing what volume of air has some particular weight. That is specific volume, marked by isosteres. Thickness lines are therefore average isosteres. In so far as they are average isentropes too, their pattern is that of the actual isentropes' contours if this is the same at all levels. The isentropic contours and isosteres, being then the same in pattern, will form no intersections, no grid. The stream lines cannot then form a grid with them either, and so must be along them. That is to say the thermal wind, whose stream lines they are, blows along them.

Now grid this whole picture with the absolute 1000 mb. contours. That makes the true or absolute wind stream lines cross the isosteres (relative contours) through their intersections with the absolute contours, i.e. with the 1000 mb. wind.

But now in so far as the air flows neither through isentropes nor through a front, isentropes follow the front and therefore move horizontally with the colder air. Thus if the isosteres carried by each isentrope are moving with the low-level cold air (1000 mb. wind) while the upper wind crosses their instantaneous positions with exactly the same velocity, the upper wind cannot in fact be crossing the moving isosteres (up hill or down hill) at all. It must be flowing *along* them, keeping to the *instantaneous* pattern of the *absolute* contours at the same time as it keeps to the *moving* thickness lines or *relative* contours, as we might have deduced anyway from V_{gs} having no divergence (to first approximation) and therefore only horizontal flow.

Isobaric analysis, introduced thus through equivalence to isentropic, will now be more easily dealt with from scratch instead. Suppose we set out to forecast the upper winds. How shall we do it? Ideally we do it in terms of pressure, because not only can we quickly deduce it therefrom with sufficient accuracy by well-known laws of hydrodynamics (even though only with almost insuperable difficulty in certain tropical or mountain regions), but pressure is most directly related to temperature and so to the air's laws of *thermo*dynamics too.

Upper winds, of course, might equally well be forecast in terms of something else, such as 'from the face of their own charts alone', as we said of our simplified charts, particularly for the tropics. In practice we use both these methods together, using hydrodynamics first, but guided by personal experience of actual charts. The technique explained in this chapter is, therefore, by no means all that there is to upper-wind forecasting. It is just based upon the hydrodynamical laws we have already learned.

First, as we know, the gradient wind at any level is shown by the horizontal isobars at that level. The forecast upper wind at 3 km., for instance is thus determined by the forecast isobars there.

Generally the isobars are moving. How do we forecast anything moving? Even if we cannot allow for the cause of motion, we can make use of the formula (well known to mathematicians as Taylor's Theorem, but fairly obvious anyway by common sense from the mere definitions) relating forecast *displacement* (l) to *time* (t), with *velocity* ($v=dl/dt$, the rate of change of displacement), *acceleration* ($a=dv/dt$, the rate of change of velocity), etc., thus:

$$l = vt + \tfrac{1}{2}at^2 + \ldots.$$

To multiply the velocity by the time, thus

$$l = vt,$$

neglecting the acceleration, etc., is an obvious first approximation. Isobars' velocity is shown by barometric *tendency*. A first approximation to forecast isobars can therefore be made from barometric tendencies as measured in practice.

That is all very well at sea-level, but not so practicable at 3 km. To forecast the winds at 3 km. we must accordingly either use 3 km. wind charts alone, or else combine the thermal and low-level patterns by gridding.

Suppose the sea-level and 3 km. pressures differ by 300 mb. That is roughly 600 lb. per sq. ft., while 3 km. are about 10,000 ft. 600 lb. are therefore the weight of the 10,000 cu. ft. of air upon 1 sq. ft. of sea surface. That tells us the average specific volume ($16\frac{2}{3}$ cu. ft. per lb.). So the relative isobars, formed by gridding the sea-level isobars with the 3 km. isobars, are simply isosteres or specific-volume lines.

Other things being equal, air that flows horizontally with no change of pressure will not change temperature either. The simple gas law relating pressure with temperature and with density tells us then that the density will not change, the specific volume is constant, and so the ISOSTERES MOVE WITH THE AIR. We deduced this before; but not in quite the same way. Now we have just said 'no change of pressure'. Moreover any wind scale used with isobars has to allow for density. Those are two complications we want to avoid. So let us start again in a new way.

What forces do we recognise on the air at any one (horizontal) *level*? Gravity has no horizontal component. That leaves only pressure gradient, geostrophic and cyclostrophic forces to determine the gradient wind. Just as any particular level is marked by a horizontal surface, so is any particular pressure marked by an *isobaric* surface. What forces act on the air in this? This time it is gravity that has some component, while pressure gradient has not. CONTOURS of this surface mark gradient wind in just the same way as horizontal isobars did. In fact they have an advantage. With them no wind scale correction for density is required. Density has already allowed for itself, so to speak, by affecting the contour spacing.

This advantage far outweighs the disadvantage of giving the wind not at (say) 3 km. but at 700 mb., i.e. not at uniform height but at uniform pressure. The wind does not reckon to vary with height all that much.

Let us try to picture isobaric surfaces just as clearly as we could picture isentropes. What makes water flow? Generally, as you know, it all boils down to *seeking its own level*. Not just at its top surface, but at any depth too. In depth, however, how is this 'seeking of its own level' to be defined? What is it, in short, that seeks to be horizontal? Simply the thin layer of water that is all under some particular pressure, that has some particular weight of water above it, say 1 ton per sq. ft. That is what seeks to be horizontal. You cannot actually see it marked out except in one special case, namely when the particular pressure is zero, that is, where no water lies above: in other words the top surface. Let us then use the word 'surface' (no longer necessarily meaning the top one) for any such cross-section under some particular pressure. With this definition you see we can simply say that water flow amounts to water surfaces at all depths seeking to lie horizontally.

That is just what we say for the air as well. Although the atmosphere has no distinct top surface but just peters out into empty space, we can certainly specify the 'surface' consisting of all the air under pressure of 1 ton per sq. ft., describing its flow as the process of seeking a horizontal level.

The total air pressure at sea-level is not actually quite as heavy as this. It is only just over 1 BAR, 1000 mb., that is about $\frac{9}{10}$ ton per sq. ft. As a cold atmosphere is, of course, heavier than a warm one, the bottom of it at sea-level is under rather more than this 1 bar pressure, so that the surface which is under just 1 bar must lie somewhere above it, like a hill. Nowhere, however, is it more than 2000 ft. up, and in hot tropical air it is found just about sea-level itself. Let us use the tropics, in fact, as an illustration.

Heating must be most intense and continuous in the tropics. The tropical atmosphere's 'ground floor', thus expanding, raises its ceiling. That makes the ceiling slope down northwards and southwards like a great flattened house roof. The $\frac{1}{2}$ *bar* (500 mb.) surface, therefore, also slopes downward out of the tropics to north and to south, making the upper air flow accordingly. Thereby removing part of the top half of the tropical atmosphere, it reduces the atmosphere's total weight or pressure felt at sea-level. Sea-level pressure, hitherto more than 1 bar, may thus be reduced to just 1 bar. That is another way of saying that the 1 *bar* (1000 mb.) surface, hitherto or elsewhere found above sea-level, is there made to sag down to actual sea-level so that it forms a trough or valley. Thus while the $\frac{1}{2}$ bar surface is sloping down out of the tropics, the 1 bar surface is sloping down into the hottest zone. And so, while the top of the atmosphere is flowing out of the tropics to try to level the $\frac{1}{2}$ bar surface into a uniform height above sea-level, the bottom of the atmosphere is flowing into the hottest region to try to level the 1 bar surface also into a uniform height.

But the air in between the $\frac{1}{2}$ bar and 1 bar surfaces, having the same weight per square foot all over the bottom whilst kept less dense over the hottest region than anywhere else, must obviously remain deepest there. So for both its bottom and top to be horizontal is impossible. They cannot be levelled out at the same time. Some slope, and therefore some air-flow, persists. That not only shows the bar surface idea but partly answers the beginner's familiar question why the air-flow does not at once fill up all the 'lows' and then stop. So we can picture, too, the 700 mb. as well as the 1000 mb. surface. Gridding their absolute height contours gives us the contours of thickness of the 300 mb. layer of air in between them. The one for 3 km., for instance, is then the isostere of specific volume about $16\frac{2}{3}$ cu. ft. per lb.

So again we see that the thickness lines are mean isosteres, while being the result of gridding absolute contours which mark V_{gs} at both levels they must also be the flow lines of the vector difference of V_{gs} values, which is the thermal wind. V_{gs} components across them are, therefore, the same at the bottom as at the top of the layer. If then the isosteres at all levels be parallel, V_{gs} components at right angles to them will be the same at all levels. Other things being equal, meanwhile, the isosteres move with the air at each level. Therefore they and their average isostere all advance with V_{gs} at any level, say V_{gs} at 1000 mb. Still with the proviso 'other things being equal', THICKNESS LINES MAY BE FORECAST BY SIMPLE ADVECTION WITH THE 1000 MB. GEOSTROPHIC WIND.

Although this is a golden rule, however, it is only a first approximation. If the isosteres at different levels are not all parallel, then the thermal wind which blows along some of them will cross others, which they will therefore advance at a different rate. So the thickness lines, being average isosteres of them all, will advance at a different rate too.

If the air is warmed or cooled as it goes along, by conduction, convection or radiation, other things are no longer equal. The main proviso is not fulfilled.

Thirdly we have assumed horizontal air-flow. Even V_{gs} is not horizontal at a front. The warm air being lighter, V_{gs} is faster and so can only escape by having an upward component. In this case, however, we can often proceed to a closer approximation. A front, remember, moves horizontally with the colder air. Also it roughly keeps constant slope—of the order of 1/50 if cold, but 1/100 if quasi-stationary or warm. Slope 1/100, for instance, means that anywhere 300 km. (or more) ahead of the sea-level front is in the cold air up to 3 km. or more, whose thickness lines move accordingly with the air. Likewise the undisturbed warmer air everywhere behind the sea-level warm front should keep constant thickness value. THE 1000–700 MB. THICKNESS LINES AT THE SEA-LEVEL FRONT AND AT 300 KM. AHEAD OF IT THEREFORE MOVE WITH THE FRONT, whether with V_{gs} or not. Then so must those in between.

As an example of upper wind forecasting based on all this, here is the writer's method:

1. Take two sheets of perspex.

2. On No. 1 trace out the charted successive positions of fronts, centres, etc., remembering to correct any inconsistencies first.

Trace them now on to the back of the perspex, and clean off the front.

The front is for things that may be revised, whereas the back is for things that will not be.

3. Laying it back on the drawn-up charts, compare the displacements from chart to chart with both the gradient and the directly measured winds (e.g. by balloon or aircraft).

Taking these into account, draw now on the front of the perspex to-morrow's positions by extrapolation from yesterday's and to-day's, allowing for anything likely to change the acceleration, such as an unstable wave development.

There are our forecast *fronts*.

4. On No. 2 (first, of course, on the front, and then on the back as you did with sheet No. 1) trace yesterday's thickness lines. Then lay it upon No. 1, both on top of *to-day's* thickness lines.

You have now superimposed

    (i) thickness lines for yesterday and to-day,

    (ii) fronts for yesterday, to-day and to-morrow.

So you can see at a glance not only the shift of the thickness lines but also any differences from the shift of the fronts, by whose extrapolation you can now sketch the thickness lines for to-morrow.

Note how much better an approximation this is than mere geostrophic advection. You have allowed for thickness lines' velocity relatively to fronts, and for fronts' velocity, acceleration and even changes of acceleration.

Better still, by including an intermediate thickness-chart among those superposed, you can see not only the shift of the lines (from yesterday's positions to to-day's) but how it has been speeding up or slowing down. In other words you allow directly for their acceleration.

5. Clean the back of No. 2 perspex. Trace your forecast thickness lines now on to the back (from the front), and then clean off the front. On the front now laid over your forecast sea-level chart you sketch (in one colour) to-morrow's absolute 1000 mb. surface contours. Gridding then in a different colour with the 1000–700 mb. thickness lines from the back of the perspex you finally draw on the front the forecast absolute 700 mb. contours with which you can use the *gradient* wind scale (allowing for curvature) to measure to-morrow's 3 km. winds.

Or you may be more interested in 500 mb. or even 300 mb. (30,000 ft. or 9 km.) winds. One level at a time, anyway. Others must then be interpolated.

This method, you see, demands neither fancy equipment nor long experience; yet it is not only quick but allows for more than many popular methods do—not least being the curvature of the forecast air-flow. It is, however, only one of many wind-forecasting methods. Others are better, in which experience counts. The subject is far bigger than has been outlined here. At least you will see how much organisation lies behind such a simple result as an upper wind forecast for aircraft navigation. Observers at thousands of places all over the world, at cities and airfields, lonely villages, outposts of Empire, islands, lighthouses, ships, icefields, mountains, jungle and desert, all observe, record and report their weather at regular hours, normally in international codes like those in this book. Their weather reports are promptly broadcast, exchanged and collected by forecasting centres all over the world to be plotted on charts so that everyone knows what is going on everywhere else.

High upper-air soundings have at the same time been made from many of these same remote stations or ships, with balloons and aircraft. The type of balloon used until recent years was normally just the small 'pilot' balloon requiring no more than one operator to fill it with hydrogen (from chemical generator or cylinder) up to a standard size, then to let it go and watch its ascent through a special theodolite, finally working out from the theodolite readings its true drift with the wind, and thus the actual wind at different heights all the way up as far as the rising balloon can be seen. That is until it bursts or is lost in cloud, in haze, in dazzling sunshine or otherwise accidentally. At night the balloon may be fitted with a small lantern containing a candle—not to be hung too close to the hydrogen! Help is welcome, but the job can be done single-handed. Rising at 500 ft. per min. such a balloon may sometimes in clear sky be watched for an hour or more, the winds being worked out as it goes up, so that after only one hour all the true upper winds up to 30,000 ft. (9 km.) can be broadcast. It demands quick work, but can nevertheless be done almost anywhere, even from a parked wagon as used by mobile units.

Although the balloon is designed to rise at a certain rate *relatively to the air* when filled to a standard size, the air is itself often rising and sinking at unknown rates relatively to the ground, and so a more refined method is for two observers to watch the balloon together through different theodolites at the ends of a measured base-line so that the rate of ascent need not be assumed at all but can in fact be worked out with the winds. There is also a single-observer method of avoiding assumption of standard rate of ascent by attaching to the balloon a 'tail' whose length is known, and whose apparent length can be measured with the theodolite so that the balloon's true distance away can be worked out. Unfortunately the tail is apt to be swung through unknown angles by odd gusts of wind.

New methods use radio instead of ordinary vision. Big balloons carrying small transmitters may be 'watched' from radio direction-finding stations at the ends of known base-lines. Pressure, temperature and humidity may be radioed too. The frequency or pitch of the signal transmitted is made to depend upon the expansion and contraction of a tiny aneroid barometer for the pressure, of a metal strip for the temperature, and of a strip or threads of sensitive materials for moisture. All contained in a small box or cylinder, they are not only cheap to produce but may be salvaged after descent by parachute when the balloon has burst. If you ever find one, you are asked to return it to the Meteorological Office, who can probably make it as good as new, and so use it again.

For measuring upper winds, however, we now prefer RADAR, for which a balloon merely carries a special reflector. The most obvious advantage of these radar and radio methods over visual ones is independence of cloudy weather, in which in the old days the upper winds above cloud were unknown or could only be guessed. A second very obvious advantage of such big balloons over small ones is that they rise both faster and further before they burst. The stratosphere can at least be reached in less than an hour. Even more recent are rocket sondes reaching 100 km. or more, such as may in the future perhaps soar to thousands of kilometres and report by direct television, superseding our charting techniques altogether.

Meanwhile we also have aircraft, with the enormous advantages of observers in person, of instruments that need not be so simple and cheap to produce (as they require no frequent replacement), and the freedom to choose not only the place and time but the track, height and methods of observation. Meteorological reconnaissance flights have various aims. In one type the aim is to fly as nearly as possible straight up over a base station at regular hours each day (the first one normally being at dawn as soon as it is daylight enough) and to record the pressure, temperature and humidity. That is no more than a radio sonde balloon might do, but then the air pilot can also explore local clouds and report their heights and what they are like to fly in, how 'bumpy' they are, what ice they deposit, and so on. He can use accurate thermometers, and fly level at standard pressures to give his thermometers time to get steady before each reading. He can of course, read wet temperature directly instead of leaving it to be worked out from the humidity such as the radio sonde gives. Above all he can see a long way.

In another type the aim is to make a long sortie at constant height (or constant pressure) on a prearranged track over sea from which there might otherwise be few reports. The base station, time and track are, if possible, always the same for a sortie with any particular name, so that all Meteorological offices know from where and when to expect the reports, and can accordingly plot them immediately on receiving them in code, even though other sorties distinguished by different names may be going on elsewhere too. Such sorties require bigger aircraft and crews, including if possible a meteorologist air observer, who is kept at least as busy as the others. Like all long flights over sea for those who are not so busy, they can be painfully monotonous when not dangerous. Descents to near sea-level may from time to time be made for recording the sea-level wind, temperature and pressure, which when radioed promptly to base station will obviously help the drawing of the latest synoptic charts. Even the rest of the flight, say at 950 mb. or about ½ km. above sea-level, is invaluable for the sea-level charts, for not only will it indicate sea-level weather and fronts, but its actual winds (as measured by drift, with visual, astro, radar or other fixes) will generally be near enough to the gradient winds of the sea-level isobars. The flight may be straight out and back, or triangular, with three 'legs', more than three legs, or even a shuttle service between different bases. Reports are made at regular intervals of time and space, i.e. at prearranged points. Certain of these prearranged points, such as the ends of each 'leg', are suitable for ascents like those at base stations. The aircraft climbs to (say) 500 mb. and may then remain at that level over the whole of one 'leg', descending at the end to 950 again for the next leg. At 500 mb. it looks down on most of the weather and also far out beyond its own track. Whilst thus able to spot, for example, far distant frontal clouds for the sea-level chart, this high flight is, of course, used for upper air charts. 500 mb. isotherms, for instance, may sometimes be sketched at once merely from the aircraft temperature readings shown on the map. Upper air temperatures (u.a.t.) for every ascent and descent are plotted on tephigrams (e.g. as many as possible together on one diagram, distinguished by different symbols or colours), from which, among other things, 1000–500 mb. thicknesses may be read along the mean d.p.t. lines with our Universal Scale. Other methods of reading thickness, too, have come into use during recent years. In this connection, the development of radar altimeters, indicating true height while ordinary altimeters really only indicate pressure, has in effect allowed thickness-line patterns to be followed by aircraft in flight, forming a navigational technique known as PRESSURE-PATTERN

FLYING. But that is another story, and so we return to meteorological reconnaissance. Plotting the thicknesses on the map at the known positions over the sea (allowing for their bodily transport or advection from the time of measurement to the standard time of the chart), and doing the same for radio-sonde ascents over land as well as at sea, with the help of the sea-level map to show where the fronts are, we can draw a map of the thickness lines. We also use every actual observation of upper wind at the chosen level, its vector difference from the known low-level wind being the thermal wind to which the thickness lines must conform. Even with observations widely scattered, there is less scope for error than might be expected, though still all too much. Again good drawing is needed.

Note that not only special meteorological flights but weather and/or temperature data from *any* aircraft at such levels will help. Reports from any aircraft at any level are useful for *some* chart, and so help others.

A type of flight that combines advantages of the ordinary type with the special type has a track planned differently each day according to such factors as:

(1) Other air-routes for which a weather-forecast is most needed;

(2) the direction from which to expect the arrival of changes of weather on these routes;

and (3) what is specially to be watched for, whether at high level or low.

The sortie is timed as shortly as possible before its use by other aircraft, its weather reports being radioed in plain language for immediate use by base or other stations, either for last-minute briefing before take-off or else for broadcast to aircraft in actual flight. Likewise weather reports in plain language or in standard code from any aircraft whatever, if radioed promptly as well as recorded for later de-briefing, can be used to help others. If you yourself fly, please help in this way. You know what to look for, and how to describe it.

2. COMPARISON OF SIMILAR SYNOPTIC SITUATIONS

You will already have noticed how often we need only quote from our earlier chapters. Now for this final week of weather reports in this book the synoptic charts, though not reproduced, are so like certain earlier ones that to describe them we need only quote our descriptions of the earlier ones with but slight alterations, and so you may see how well we could forecast the weather by mere comparison. You might well expect it to be much quicker than having to work it all out from scratch.

Our last picture was for the morning of **14 December**. Our first report now, you notice, is of heavy rain (and even a heavy squally shower of rain and hail) in the evening. That is due to the cold front, so that next morning, **15 December**, we find ourselves in cold, clear polar air.

'Meanwhile', as on 25 October, 'the liner *Montrose* steaming westward at about 54 N...', this time at 25 W instead of 35 W, has reported

(i) fresh west wind with hail showers, temperature 40° F., and bar. almost steady in the polar air;

(ii) south-south-east gale with temperature 46° F.;

(iii) squally west-north-west gale with temperature 43° F., and bar. 21 mb. lower; and then (at 18 G.M.T., 15 December)

(iv) the same squally west-north-west gale with showers (of snow this time), temperature 40° F., and barometric tendency now no longer falling but rising 10 mb. per 3 hr.

'The next wave depression, which these observations have clearly located ...arrives here to-day'—or rather to-night as you see by the drizzle, the clouds, the wind and the temperature.

'Our front's arrival here...is exactly confirmed by the official charts, although they are not reproduced in this book. You will find we can manage quite well now just with descriptions of the original charts, or rather of just the features that matter to us.'

To-day's shows our front clearly marked by the 'wind, barometric tendency, clouds and rain. From the centre of the depression this front is linked to yesterday's which is now over Europe...to the Arctic Ocean. Two arctic stations provide a fix. One of them is Bear Island...' where a south-east breeze is replaced by a south-south-west gale with rain.

'That shows the front all right. The other is Jan Mayen, where with a north-easterly gale the bar. has fallen to 964 mb. That is so low as to be evidently about the old low-pressure centre.' Actually it is better marked by rapid wind change from east-north-east to west-south-west as bar. reaches 952 mb. Again it is due north from the Shetland Isles or rather the Faroe Islands area where it appeared on the previous evening.

'The new depression centre already shown by *Montrose*...to have been about 54 N' at longitude 25 W on the morning of 15 December, 'is next revealed' over the Faroe Islands (24 hours later) 'where bar. reaches

Place no.	Date 1936	Time G.M.T.	Weather letters	Cloud				Vis. V.	Wind		Temp.	
				T.	L.	M.	H.		D.	F.	F.	C.
01	14.12	1920	o RR q	8	7	–	–	6	S'W	5	47.6	09
01	14.12	1950	o rr	8	0	2	–	6	S'W	3	47.1	08
01	14.12	2030	o RR	8	0	2	–	6	SSW	3	46.9	08
01	14.12	2100	o rr	8	7	–	–	6	SSW	3	46.4	08
01	14.12	2200	o PHR	8	9	–	–	6	SW'S	6		
01	14.12	2203	o RR q	8	9	–	–	6	SW	6		
01	15.12	0710	b	1	5	0	0	6	WSW	3	36.7	03
18	15.12	1100	b	0	0	0	0	6	WSW	3	41.8	05
01	15.12	1845	b	1	0	0	4	6	WSW	3	38.9	04
01	15.12	1950	o	8	0	7	7	6	SW	3	39.9	04
01	15.12	2030	o q	8	0	2	–	6	SSW	4	40.9	05
01	15.12	2215	o d/D	8	–	2	–	7	SSW	4	42.3	06
01	16.12	0745	o r_0r_0 q	8	7	–	–	6	SSW	6	45.6	08
18	16.12	1100	bc	4	3	7	2	6	SW	3	49.4	10
01	16.12	1340	b	1	1	0	8	6	WSW	4	47.5	09
01	16.12	1420	c	6	8	3	–	6	WSW	4	47.9	09
01	16.12	1450	c pr	7	3	3	–	6	W	3	47.5	09
01	16.12	1520	c	7	8	7	–	6	W'S	3	46.0	08
01	16.12	1600	b	1	0	3	0	6	WSW	3	43.6	06
01	16.12	1620	b	1	0	7	0	6	WSW	3	42.9	06
01	16.12	2200	b	0	0	0	0	7	S	3	41.2	05
01	17.12	0745	c	8	7	7	–	6	SSE	3	44.1	07
18	17.12	1100	o r_0r_0 q	8	7	–	–	6	S	6	48.9	09
18	17.12	1230	o d_0	8	7	–	–	6	SSW	5		
01	17.12	1800	o r_0r_0	8	7	–	–	6	SW	5	52.2	11
01	17.12	1830	o rr	8	7	–	–	6	SW	6	52.0	11
01	17.12	2145	o r_0r_0 q	8	7	–	–	7	SW	6	53.2	12
01	18.12	0750	o	8	5	–	–	7	SW	6	51.8	11
01	18.12	0800	o r_0r_0 q	8	7	–	–	7	SW	7	51.8	11
18	18.12	1100	o d q	8	7	–	–	7	SW	7	53.2	12
01	18.12	1730	o d/R	8	7	–	–	6	SW	3	49.7	10
01	18.12	1800	o d	8	7	1	–	6	SW	3		
01	18.12	2015	o rr	8	7	–	–	6	SW	3		
01	18.12	2130	o	8	7	2	–	6	SW'W	3	49.0	09
01	19.12	0730	b	0	0	0	0	6	W'S	3	40.1	05
01	19.12	1200	bc	3	0	0	9	6	W	3	40.4	05
01	19.12	2250	o	8	5	1	7	7	SW	4	44.0	07
01	20.12	0945	bc	3	1	3	0	6	SW	4	44.0	07
01	20.12	1230	bc	3	2	0	0	6	SW	5	47.1	08
01	20.12	1950	c	7	4	–	–	6	SW	5	45.6	08
01	21.12	0740	c/b	6	5	0	0	6	SW	3	43.6	06
01	21.12	1310	c	7	5	0	0	7	SSW	3	48.3	09
01	21.12	1650	b	0	0	0	0	6	S'W	3	42.3	06
01	21.12	1920	o	8	5	–	–	6	S'W	2	41.1	05
01	21.12	2215	o	8	5	–	–	6	S'W	3	42.7	06
01	22.12	0800	o	8	7	–	–	6	SSW	3	44.9	07
01	22.12	0920	b	1	1	3	0	6	SW	3	44.8	07
19	22.12	1400	c	7	8	3	0	6	SW	3		
01	22.12	1830	o	8	5	–	–	6	SW	4	50.0	10
01	22.12	2000	o r_0r_0 > RR	8	7	–	–	5	SW	4	44.8	07

minimum' (of about 943 mb. this time) 'while the wind with force 6 or 7 (on our Beaufort scale) backs right round' from south-east to north-westerly.

Meanwhile as *Montrose* bar. minimum was 975 mb., whereas the Faroe Islands report 943, 'the "low" has deepened 32 mb. in the 24 hours. That makes average tendency "deepening term" —4 mb. per 3 hr.'

Next compare 27 October. This time it is not 'the liner *Ascania*, steaming west about 53 N 24 W at midnight' but the liner *Transylvania* steaming west about 53 N 35 W at midday that reports successively

    (i) in the high pressure ridge, bar. rising, wind west by north (force 7), sky half clouded with small cumulus;

    (ii) at midday a few hours later, bar. falling (7 mb. per 3 hr.), wind south-south-east (force 4), and sky overcast with ragged low clouds of bad weather;

    (iii) a few hours later, bar. falling 1·4 mb. per 3 hr., wind veered (to west, force 4) with broken clouds and showers.

It is almost word for word as it was on 27 October. The only important difference is that it is so much colder that the warm-air sectors of the depressions are not now met by these ships, being further south all the time. The next one, however, comes over us as you see marked on **17 December** by rain with a south-west wind and high temperature which last over **18 December** too.

3. PRACTICAL FORECASTING

What, then, of the morrow, 19 December and the next few days? We must expect our warm air soon to be swept away by a cold front from the south-west or west and then followed by polar air.

What will that mean? As on 7 November again, 'the general picture is slow to change, a general forecast for England simply being of strong south-westerly winds with showers and otherwise good visibility'.

That reminds us next that the forecaster seeks minor fronts or troughs in the air stream. 'A minor front always does something. In fact it is often the only way to explain or anticipate certain remarkable changes of weather from what you would expect at first sight of the picture....A minor front in fine weather might be marked just by an altocumulus cloud layer or by sudden haze thickening. In fair weather it might be marked by a bank of Cb among Cu clouds. In cloudy weather it might be marked by a short deterioration with low stratus clouds and drizzle....All these may only be signs of a minor front; but these very signs themselves, sudden banks of cloud or deteriorations in visibility, all matter a lot to those who fly, to whom any failure of our anticipation is a serious error that may even be fatal.' In any case we have to forecast the height and amount of the clouds according to all our laws of thermodynamics with u.a.t. measured or estimated as far as possible up-wind beforehand, allowing for subsidence, fronts, etc., according to the air's hydrodynamics too. Also we have to forecast the winds in detail.

Difficult though it may be, this is only the pleasantly scientific half of a forecaster's work. On a 12 hr. night watch, for instance, it ought not to take up more than 6 hr.; for, like your daily newspaper, it has then to be written or typed or printed, illustrated, duplicated (often all by hand), telephoned, teletyped, broadcast or published in time for all who require it for their morning work, whether by air, by land or by sea. Allow, too, for time to 'brief' crews, to answer 'spot' queries from all and sundry, and above all to go on watching the weather reports to see that the story remains correct, or otherwise to amend it all at short notice, and you will begin to see what the work really is. Working it out, in short, is not even half the battle. If anything wears a forecaster out, more than the battle of working it out against fatigue and distractions day after day, or night after night, it is the battle of putting it out against time. So we come to the gentle art of putting it all across. Do you buy weather forecasts, or sell them? Are you a taxpayer with a State Meteorological Service? Then obviously you buy them. Do you fly with an airline paying fees for services at state-owned airfields? Again you see you are buying them. Do you get your money's worth? Are you, on the other hand, in the Meteorological Service yourself? Then you sell your knowledge. How much do you sell? As a monopoly selling experts' forecasts, whether your own or others', you will always have customers, making it all too easy for you to think you might get away with poor salesmanship. Naturally you are trusted not to think so. The official forecast may be wrong, but is less often wrong than anyone else's is likely to be; so naturally your customers risk it, assuming that they are at least going to learn from you how it will affect them. Then it is up to you, by drawing, writing or talking to them, to make them foresee their own weather as clearly as it is foreseen by you, or as it is foreseen by experts, whether you yourself are yet expert or not. You must at least understand it yourself, even if you have not had to work it all out.

Drawing it, whether in map or vertical cross-section form, is an art, which some otherwise excellent meteorologists have not mastered. There is as much art in it as in a poster. Too many posters, you know, try to say far too much. How often do you actually read every word of long-winded advertisements? Few even catch your eye enough to make you start. Most posters, so to speak, go in at one eye and out at the other. What really catches your eye is something unusual, bold and brief.

So let it be with your weather pictures, particularly with pictorial cross-sections for those who fly. Space out the clouds so that whilst shown in the right zones in the picture they also look something like natural shape, though simplified as you would like them to be in a poster. Try to avoid distorting their shapes too much, even though an inch on your paper may have to represent 100 km. horizontally but only 1 km. vertically. Those who fly may be used to it and so allow for it as a convention; but why not give them one thing less to allow for? Have they not already enough in their cockpits?

Don't, on the other hand, be unconventional in other ways without due warning. Just make your drawings appeal to the eye without being misleading.

Simplify horizontal (synoptic) charts too, like ours for example with 'wind flow lines,

(a) drawn with the wind directions, and spaced apart inversely as wind speed;

(b) spaced as widely as possible, provided that they show all the main air streams;

(c) always meeting at fronts in such ways as air streams must;

(d) each to be traced if possible right across the map;

(e) always clearly accounting for all the air.'

Showing weather types too, if not fronts and air masses, 'this kind of weather map is suggested for showing clearly to air crews and laymen not only the past but also the forecast weather'.

Remember, too, 'weather inference with synoptic charts is never meant to be perfect, but merely to be as full and as nearly true a picture as can be drawn in time to keep as far ahead of the weather as is required. There is psychology in it. It depends upon who wants to know the weather, how reliably they need to know it, in how much detail, and how long ahead. Supply depends on demand. The picture, however, must always be served up quickly and clearly. That is what calls for artistic sense as well as a sense of proportion. You need a sense of proportion to tell you what matters and what does not, and so to know where to smooth out irregularities that do not matter....Nature is complex; but Nature when averaged suitably for our purposes on a map always shows a smoothness of pattern that to an artist often has beauty....To be able to "hit off" the isobars' spacing and curvature with the very first bold quick strokes of a pencil, calling for little or no india-rubber, is an artistic gift that a forecaster should appreciate. Not only is the finished work a thing of beauty, but also it saves much vital time in the first one or two hours when the chart is used for an up-to-the-minute forecast.'

Graphs, too, even if not for public display, should be drawn almost as if they were; for when you yourself see a technical graph, do you not prefer it with bold clear legends and lines than all cluttered up with only half-explained figures? It is the art of window dressing. The goods may be there all the time, but to make them stand out is the art. For your enterprise and imagination, even in planning the coldest scientific graph, there is far more scope than you think.

Consider now *writing* a forecast. If ever you have to work one out from synoptic weather reports, the writer would strongly advise you first to PLOT FOR YOURSELF, if you can, the latest synoptic chart you will use. Not necessarily any earlier ones, nor necessarily even the latest *main* chart (e.g. the midnight chart for a 4 a.m. forecast). *Their* plotting can probably be left to an assistant while you are engaged on post-mortem analysis of the previous day as a guide to the following day. But if you can possibly spare the time, take over from your assistant to plot for yourself, say, the 3 a.m. chart to be taken into account for your 4 a.m. forecast. Not even all of it; only as much as you need for this forecast. Your assistant can do the rest afterwards. Above all, let it not merely go in at one eye and out at the other. TAKE IT IN as you go along.

By this method alone you have it not only analysed but all in your head as soon as finished, as soon as plotted, that is (or should be) as soon as the data are in, which in turn may be within half an hour of their observation. You may even spot fronts etc. before you have finished. A mere glance down your coded list of weather reports, like the glance of any experienced eye down a page of reports in this book, may suffice. Here in London, for instance, weather letters 'rr' would imply a front. Experience teaches; but speed with it counts above all. YOU ARE ALWAYS EXPECTED TO TAKE ALL YOUR LATEST AVAILABLE DATA INTO ACCOUNT. To give yourself plenty of time by writing a forecast too early is either a waste of that time (if you have to amend and rewrite after all) or else just one of those things that is not done, and may even be dangerous, ignoring later data that might have been used. That is why much of the work is done against time, and why it can be a great mental strain. Do well in advance all the donkey work needed, such as the heading and setting out of your forecast forms, carbon papers, folders etc.: but not the forecasts themselves.

If the drawing of the forecast weather is art, then the writing of it must be literature, or at least something like good journalism. As such, it should be (*a*) concisely set out with conventional headings, (*b*) clear at the very first reading, and (*c*) free from catch-penny phrases or otherwise unconventional language. Its language, in fact, is highly conventional, sometimes too much so. The writer is all in favour of expressing himself like a human being, not in technical language. That, however, is all very well in talking but not in writing.

No forecast is ever perfectly certain. No useful one, anyway. Its uncertainty may be expressed in various ways. One way is with exact words in vague phrases. Another is with exact phrases but vague words. The latter way may sound more definite and concise, but is less widely used, less well known and therefore less well liked. Moreover it does not work with numbers in any of our most common units like miles, kilometres, knots or oktas. That is really due to our well-known scientific convention that when any figure is given, it is supposed to be accurate down to the smallest unit mentioned. If, for instance, we say '5·50', we mean it to the nearest hundredth: if '5·5', then to the nearest tenth: if '5½', then only to the nearest half. If '5', it is only meant to the nearest unit. But even a unit is often more accurate than a weather forecast can stand. A slight breeze, No. 2 on the Beaufort wind scale, is 4–6 knots. Though true as an average, '5' implies too great a precision. The wind itself does not know its own speed so precisely. To fluctuate is in its nature. So we should actually say '4–6 knots'. That, in fact, is how forecasts come to contain not single figures but *ranges*, like 'wind south to south-west, 4–6 knots; visibility 6–12 miles...'.

You notice that like the wind speed, the visibility range corresponds to a single definite figure (in this case '7' in our book, meaning 'visibility 10–20 km.') in the international code. That is how many stock phrases have been introduced, such as:

'1000–2000 yd.'   (approx. vis. No. 4)

'2000–4000 yd.'   (approx. vis. No. 5)

'600–1000 ft.'   (cloud height No. 3)

'1000–2000 ft.'   (cloud height No. 4)
'2000–3000 ft.'   (cloud height No. 5)
'Fine weather'   (letter 'b' or cloud-amount 0–2 oktas)
'Fair weather'   (letter 'bc' or cloud-amount 3–6 oktas)
'Cloudy weather'   (letter 'c' or cloud amount 7–8 oktas)
'Intermittent rain'   (letter 'ir' as explained in Chapter 11)
'Occasional rain'   (letter 'ir' as explained in Chapter 11)
'Showers'   (letter 'pr' as explained in Chapter 11)

The only way to avoid such terms, in fact, is to use the code numbers themselves. Meteorologists do so between one another until by nature they not only talk but think directly in code. For still greater brevity the writer suggests the use of his weather-type numbers too.

Though these stock phrases are simple translations of code numbers, do not just use them on principle. USE YOUR IMAGINATION, AND PICTURE THE WEATHER BEFORE YOU DESCRIBE IT. Think how stratus on hills will appear to a pilot above it, how varied are the top levels of Cu and Cb clouds to pilots above or among them, or how their bases come down in a shower—not down to a uniform level of 600 ft. as the coded forecast in figures is apt to suggest, but just here and there. Picture, too, not only the weather itself but the land, sea, snow, mountains, forests, fronts, isentropes, air-flow, all the weather's causes. Whether exact in words whilst vague in phrases, or exact in phrases whilst vague in words, the vagueness of description should just correspond, no more and no less, to that of the forecast picture. If, for example, a front is timed to arrive in the afternoon, but no more precisely than that, then you say it is due 'in the afternoon' or 'between about 12 and 18 G.M.T.' or whatever it is. 6 p.m., by the way, is normally called 1800 hr., but if it is just to the nearest hour we should rather just call it 18 hr. Convention again. If you can double the accuracy, or in other words halve the probable error, then you may say for example 'between about 12 and 15 hr.' or 'early afternoon' or 'soon after midday', midday being taken as roughly 11–13 hr. Remember always to say whether times are G.M.T. (Greenwich Mean Time) or B.S.T. (British Summer Time) or E.S.T. (Eastern Standard Time) or whatever it is. G.M.T. is denoted by 'Z'; 1 hr. later than G.M.T. is 'A'; 2 hr. later than G.M.T. is 'B', and so on. Thus you can always choose between vague terms exactly put (like 'soon after midday') or exact terms vaguely put (like 'between about 1200 and 1500 Z').

How else may uncertainty be expressed?

One way is to classify CONFIDENCE, to say whether average, high or low, whether in doubt about actual safety (e.g. in a risk of thunderstorms) or only of navigation (e.g. with uncertain winds). Another way is to state the possibility, probability, chance or risk of this or that or the other. Notoriously open to abuse, such words when correctly used will express not only what the forecaster has to say, but what the man in the street has to know—whether, for example, to take a raincoat or risk going without it.

If an event's probability is appreciable at all, then we say the event is *possible*. There is a *chance* of it. If the event is to be unpleasant, this chance is a *risk*. If probability is more than about 50%, the event then being more likely than not, we say it is not merely possible but probable. But remember that 'the quest of anything near perfection in weather prediction is a mere wild-goose chase', so that you must be content with a risk. If a forecaster mentions a risk, you have something to work on. By no recognised scientific method could you really have more. Science recognises that, at any rate in a certain sense, nothing is quite impossible. So must you. If the forecast mentions the risk of something that does not come off after all, the forecast may well be correct—as far as it goes. If it mentions a risk of showers, for instance, which miss you after all, you may yet have seen them nearby, and so have seen for yourself that the forecast was good enough. That, however, is how words like 'risk' are open to so much abuse. Anyone could forecast correctly by mentioning enough possibilities. Worthwhile forecasting does not altogether avoid them but compromises between being useless through undue brevity and being useless through undue precision resulting in errors.

Brevity is important. Like uncertainty, it may demand vagueness of language. That cramps your style! Just try to describe, in a sentence or two, the whole country's actual weather over 24 hr.! No forecast should be rashly paraphrased or cut about by a layman. It is quite hard enough for the forecaster to express himself in limited space as it is, without the whole sense of his forecast being lost or misconstrued by unauthorised omission of what may turn out to have been the most important words. Forecasts sent by telegram, or by radio-signal, need utmost brevity. To make them perfectly clear, with the fewest possible words and punctuation marks, is quite an art, sometimes taking almost as long as the rest of the work put together. If ever you do it, be sure to allow yourself time, preparing the message form beforehand as far as you can. Think of the signaller who has to send it in Morse code without time to stop to see what it is about; so WRITE IN BLOCK LETTERS AND DO NOT PUT ANYTHING THAT CANNOT BE SENT

IN MORSE. That is easy, of course, if your message is coded already in figures (normally grouped in fives), but not so easy if in plain language.

So much for the *writing* of forecasts. Now what about explaining them verbally? That is normally either by personal briefing, such as to air crew, or possibly 'over the air'. It also includes the answers to ceaseless distracting personal or telephoned enquiries received at a busy Meteorological office from here, there and everywhere, ranging from Timbuctoo's average annual rainfall to the holiday weather prospects for Scotland the week after next. A common enquiry is the immediate local weather forecast, or particular parts of it, such as for example the cloud amount or upper winds. CARRY THESE IN YOUR HEAD (not forgetting, however, to revise them as the day goes on), so that you can always rattle them off without having to keep your customers waiting while you look them up. It not only saves time but may enhance your customers' confidence in you. Above all, keep an eye on what is going on all the time, both inside your office and out. Just an eye is enough; you may often be up to the other in work! If it starts to rain, you must not need someone else to point it out to you. As a forecaster anyway you should glance out of the window to see what is coming before it arrives, calmly warning your principal customers (and any others for whom you have time, in priority order); for whatever you actually see in the sky not far up-wind of yourself, however uncertain previously, will probably pass over you in the end. Never mind if it is not what you forecast before. Better a bit late than never, as long as it is not too late. You will be forgiven. It is just a case of taking your latest data into account, as you always should. Only if you fail to do that will you not be forgiven. Never lazily quote any ready-made forecast unless you have only just made it, or else have checked that it is still expected in the light of new data, including what you can see in the sky outside. The check need not be elaborate. With experience you will know almost by instinct whether what the old forecast said is now wrong.

Likewise let no assistant quote any ready-made forecast unless satisfied that it is still up to date. The writer believes in telling his own assistants what is expected and why, for what to look out, and so how to make themselves even more helpful than ever. Are you one? Then if you have read this book, perhaps you realise more what the subject is all about, and how you can help not only a busy forecaster but customers too, particularly by carrying latest facts and figures in your head, so that you can quote them promptly. Not always need they be very precise. Except perhaps for bar. on aircraft landing, round figures will often suffice. Perhaps your trickiest

thing to estimate is the height of clouds, for which not only should you
make use of any local air-pilots' reports but you would do well to ask your
forecaster's advice; for even if in estimation of cloud height by eye he is no
more experienced than you, yet he may have indirect evidence from charts,
or particularly from tephigrams. Remember the useful rule that the height
of base of convection (Cu or Cb) clouds is often about 220 ft. multiplied
by the depression (in ° F.) of dew-point below dry-bulb temperature (at
screen thermometer level).

All this advice about answers to verbal enquiries also applies to briefing.
For those who fly, for example, a route forecast is written or typed on
printed forms (one set per aircraft), which have to be ready for pilots and
navigators as soon as they come to be briefed. That is usually an hour or
more before take-off. Unless you have not been warned of their flight at
all, you should not keep them waiting. You should have arranged beforehand to know what to write in sufficient time to get it all written, illustrated
and duplicated if necessary, before due for issue.

If, therefore, others have had to work out the forecast for you to copy,
they must have done it earlier still. For a 7 a.m. take-off, for instance, their
time of origin (t.o.o.) might have to be 4 a.m. So they would have based it
upon the 3 a.m. chart. They would then have to write it down to be read,
telephoned, teletyped or otherwise signalled to you, allowing plenty of
time for actual sending, receiving and perhaps also decoding, quite apart
from delays by other traffic and technical hitches. You might be lucky to
get it by 5 a.m. By the time you have checked it, argued if necessary with
its originators (who are probably up to the eyes in work now for others),
written or typed it out, drawn the maps and cross-section pictures to go
with it, perhaps had to make many copies by means of hectograph jelly or
stencil duplicator, and done the same with several other forecasts for
different routes, you will have done well to be ready for all the crews
by 6 a.m. That is your time of issue (t.o.i.). By this time the forecast
might easily need alteration, particularly in what it says about 'take-off'
conditions, e.g. at dawn with a risk of fog patches. Take-off might accordingly
have to be postponed for an hour or two. That would alter the time of day
for the whole of the forecast, so that the rest might have to be altered too.
That is why whoever originates a forecast—whoever first words it—is well
advised not to be too precise but so to word it as to cover such changes. If,
for example, he foresees any serious risk of fog delaying take-off, he might
simply forecast what to expect whenever it clears.

The actual times, however, must be filled in at least by the t.o.i. Whatever is forecast on paper is taken to be for the places and times that appear on the paper too. If there is a crash, the part (if any) played by your weather forecast will be judged by what is found on your file copy of the actual paper issued. If take-off is postponed you should therefore remember to re-head the paper with the new times, before handing it over. And re-head your file copy too.

Further points for beginners in briefing are:

1. Will the flight be at high or low level, above low clouds or below them? Your written forecast is for either, but verbal briefing for high-level flight need not dwell on what happens far below it, except where the aircraft will have to climb or descend through it. Neither need you dwell upon high-cloud details for take-off, landing and low-level flight, except perhaps how much sun, moon or stars will be covered up. Use your imagination.

2. Mention first whatever will matter most. It might be a front, or some fog, or head wind so strong that the whole plan of flight needs alteration. (Flight planning, by the way, often needs wind forecast before the rest of the weather forecast.) Show how to avoid these snags. If it is impossible, say so. The flight might then be cancelled or, rather, postponed until you expect better weather. If, on the other hand, all is well, then just begin by saying what *type* of weather it is.

3. Allow for the crew's experience. A crew experienced on some particular route, after being shown the chart and told of anything special, need only hear whether the rest is normal there for the time of year or the time of day or for your particular type of picture. They can read your details at leisure, or such little leisure as they may have, whereas for new crews you should explain more.

4. Quietly show that you not only know your job but have some idea of theirs. You will not make their decisions for them, but you will certainly guide them. Guide them, too, the same way as your seniors guide their commanders, all with the same weather forecast. Let it not be felt that the Meteorological Service is divided against itself. It must agree sometimes to disagree and so confess its uncertainty, but still it should be 'what Met. says', not just 'what So-and-so says', except in so far as So-and-so is an outstanding authority with whom other meteorologists are prepared to agree.

5. Distinguish weather *forecasts* from weather *reports*. If asked for a weather report when a forecast seems to be wanted, make sure which is

meant. 'Report' should mean only of present or past conditions. If of actual observations (as opposed to any uncertain inference from them) it may be called an *actual*. Some fighter-pilots, and others who either do not want to know long ahead or else perhaps have not enough faith in meteorological forecasts, ask for 'actuals' only. Give them promptly, and then if the weather is changing seriously you might add what you think it is now, or what it soon will be.

6. Always arrange not only to give your forecasts but to receive reports from all whom you brief, and from others, too, who come in to any station of yours. Help them to learn what to tell you, what to look out for and how to describe it, whether in code or plain language. 'Weather reports in plain language or in standard code from any aircraft whatever, if radioed promptly as well as recorded for later de-briefing, can be used to help others.' For any one flight, de-briefing is the forecaster's last but not least job. Like briefing, it is a bit of an art. You can use it directly for others about to go the same way, allowing, however, for weather changing by then.

So your work goes on, procedure the same every time, but with different weather. Even if constant where *you* are, the weather varies not only from day to day but from hour to hour at least over some of the routes and regions for which you forecast. Many people never think of that. In the tropics, for instance, they used to think (and perhaps still think) that all a forecaster needs is a calendar, just to tell him the day when tradition says that the weather changes from fine to rainy, or from rainy to fine. Nowhere in the world (except perhaps in the desert, and not always even there) is weather forecasting really easier, on the whole, than you have seen for yourself in this book.

So with this book not only may the general reader learn something of what lies behind the Meteorological Office forecasts and inferences and how far he may make his own from *Daily Weather Reports*, but other beginners in the Meteorological Office itself who already know how to make weather charts will have a better idea how to use them. Many books for beginners in meteorology have dealt with *some* of the things referred to here, as well as with other things hardly discussed here at all. This book's other topics, on the other hand, are normally only discussed in technical literature. Although no specific recommendations are made here about what else to read, the distinction between what is covered by 'popular' books and what is only covered by more technical works is worth bearing in mind.

INDEX

Printed in the United States
By Bookmasters